THE GOOD AND THE BEAUTIFUL
NATURE READER
INSECTS & ARACHNIDS

written by
MARY ADRIAN,
MARION W. MARCHER, *and*
PAUL MCCUTCHEON SEARS

Original illustrations by Barbara Latham, Ralph Ray, and Glen Rounds
Cover design by Phillip Colhouer
© 2020 Jenny Phillips
goodandbeautiful.com

CONTENTS

Firefly .4

Garden Spider 41

Honeybee 79

Monarch Butterfly 130

FLY

by PAUL McCUTCHEON SEARS

illustrated by Glen Rounds

Originally published in 1956

FOREWORD

Fireflies belong to a family of beetles named *Lampyridae*—a family with more than 1,500 species, about 60 of which are found in the United States. They differ principally in the way they flash their lights, which are mating signals.

In the western part of this country, there are fireflies that do not produce light as adults, although the larvae often glow. These non-flashers are active by day instead of night.

This book is about a light-signaling firefly (of the genus *Photuris*) whose life story is much like that of our other fireflies. Only the final few weeks of its two-year life span are spent in the winged, flashing adult beetle stage. Most of the time, it leads a hidden life in a form that does not even resemble the adult.

The "living light" of fireflies has been studied

by many scientists. Much has been learned about its chemical nature, but the way fireflies turn on their light is not completely understood.

Firefly light is sometimes called "cold light" because the amount of heat it produces is so slight. For example, the flashing firefly does not feel warm to the hand.

I wish to thank Mr. Frank A. McDermott of Wilmington, Delaware, for checking the manuscript for accuracy. Mr. McDermott, a physical scientist, has worked on chemical problems of living light and has published papers on fireflies in professional journals for many years. In addition, Dr. C. Clayton Hoff at the University of New Mexico gave valuable help in locating materials concerning the habitat of the firefly in this book.

Paul McCutcheon Sears
Albuquerque, New Mexico

NIGHTFALL ON THE MEADOW

The long summer twilight was slowly turning to night.

Far overhead, nighthawks hunted for high-flying insects. Down closer to the trees, bats fluttered after lower-flying insects. In the still air above the meadow, thousands of fireflies were flashing.

On the ground a skunk pounced on beetles before they could fly. Mice nibbled on grass stems. Toads hopped after moving insects.

Among the grass blades, smaller animals were hungry, too. Spiders waited in webs. Crickets chewed tender plants.

Down among the grass roots and in the top layer of soil, even smaller animals were eating. These were mites and springtails and bristletails, feeding on dead leaves and roots.

Still farther down in the dark soil, worms were crawling—earthworms and cutworms and wireworms of many sizes.

Twilight was over. The only light above the meadow came from the flashing fireflies and,

more faintly, from stars in the sky. Now that it was dark, the firefly lights shone more brightly.

One firefly walked on the ground, turning and stopping and blundering ahead again to get around the big grass blades. She flashed her light dimly as she walked along.

At the end of her body was a long egg-laying tube. She stopped in the darkness and pushed the tube into the ground. Then she pushed an egg from her body down through the tube, into the loose soil. She left the egg there, hidden below the ground, and blundered on through the dark jungle of grass.

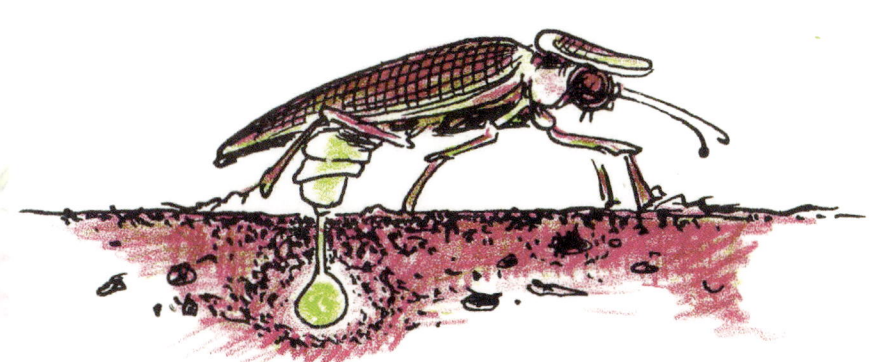

JUST BELOW THE GROUND

The round yellow firefly egg stuck to the bits of soil around it. The loose soil let in air, but it would keep away the heat and light of the sun.

Inside, the egg was full of yolk, except in one part where a tiny worm-like animal was growing. It was using the yolk for food.

In three weeks the yolk was almost gone. The tiny animal nearly filled the egg and lay curled up with its tail tucked under its stomach.

Underneath its body, near the tail, two small round lights began to glow. They shone through the covering of the egg, glowing with a faint yellow light in the dark soil.

In another week the egg hatched. The tiny animal came out with the two round lights

shining more brightly. It was not yet a firefly, but a glowworm. It had weak legs and eyes, short antennae, and strong jaws, or mandibles (MAN-dih-bls).

Glowworm had hatched out in a clump of soil held together by grass roots, like twisted ropes. All around her was the dark world of the soil, a world of caves and tunnels and passages.

She scratched with her six thin legs and turned her body in the soil until she opened the way to a tunnel made by some other tiny animal—a tunnel just her size.

Glowworm: actual size is less than 1/8 of an inch

Glowworm's small legs could hardly drag her long body through all the twists and turns and low places in the tunnel. But she had a way of making the walking easier.

At the end of her body was an odd tail like a fringe of little strings. Into this tail she pumped blood from her body. This made the fringes stand out like a row of small bristly feet. The fringe foot pushed from behind and helped her walk.

Her yellow lights did not help her weak eyes to see down here. So she felt her way along the dark tunnel with her antennae. She came to a

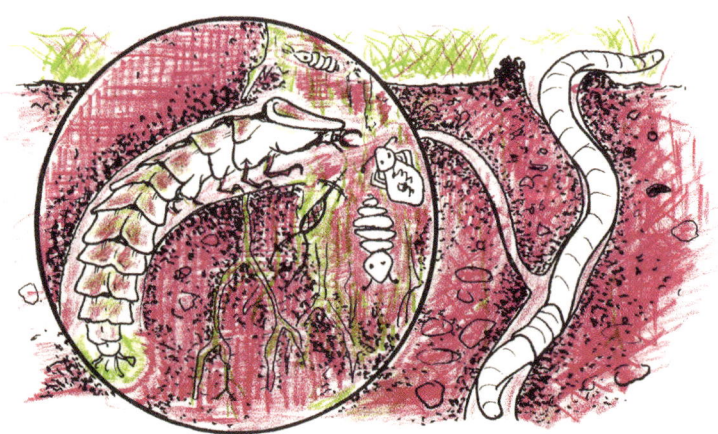

glowworm and soil animals

pebble that seemed to fill the tunnel. She felt along the edges of the pebble until she found a way around it.

Beyond the pebble she came to a large dead root lying along one side of the tunnel. Hundreds of mites and springtails and bristletails clung to the root, eating its soft, rotted insides. When they finished it, there would be more room in the soil for the roots of live plants to push down from above.

Glowworm did not try to eat the root. She was a meat eater. She caught some of the tiny animals for her first meal.

Then she crawled on. The narrow, twisting tunnel opened into a much larger tunnel with smooth sides. It led upward. Glowworm had come to an earthworm's tunnel. She crawled up, out of the ground.

It was night. Glowworm felt her way in the darkness, hunting for food. She felt everything she came to with her quivering antennae, to learn what each new thing was.

When day came, she did not like the light. So she crawled down again into the top layer of soil and found a dark hiding place.

After that, she hid quietly in the soil by day, and at night she hunted above ground.

During the summer she grew larger. When frost came, she dug out a resting place for the winter under a stone. As she became quiet, the faint lights on her tail dimmed, but they did not quite go out.

one kind of springtail

GLOWWORM, THE MEAT EATER

When spring came, the soil grew warm. Glowworm stirred in her resting place, and the dim glow of her lights brightened.

The bristletails and the springtails were coming out of their resting places deep in the soil. Although they were insects—the smallest insects of all—they had no wings because they spent much time underground. They crawled through the passageways. But whenever there was room, the springtails jumped along by flipping their tails.

another kind of springtail—jumping

One night Glowworm came out from under the stone. She crawled among the grass blades on her thin legs, pushing her clumsy body along with her fringe foot. She stuck her small head out from under the shell of her body. Her lights shone faintly on the ground, but they did not help her find her way.

She ran into something much bigger than herself, which made her stop. It was a small earthworm stretched out on the ground.

She tapped the worm two or three times

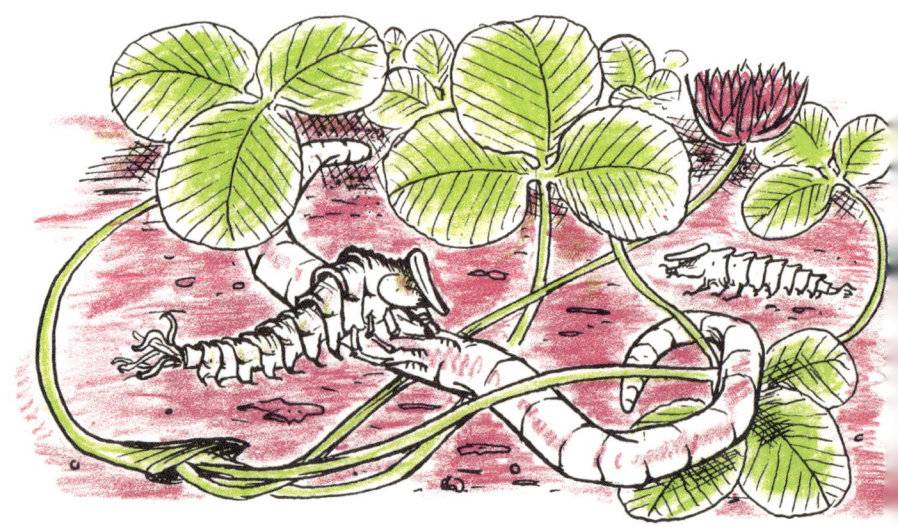

with her antennae. Then she bit him with her curved, pointed mandibles. The earthworm twisted and wriggled, but Glowworm held on. She tried to grip a plant stem with her fringe foot, but the worm dragged her away.

After twisting a minute or two, the earthworm quieted down. Glowworm took her mandibles out and felt for another place to bite him. This time he twisted again, but not so strongly because Glowworm sent a poison into him. She squeezed it out through openings in the sharp ends of her mandibles.

Other glowworms came, feeling their way in the dark. They touched the injured worm. They bit it too, and in a few minutes, it was dead. The poison was a digestive juice that turned the worm's body into a liquid.

Soon five glowworms were sucking in the liquid. Their tiny heads were almost buried in the earthworm's body. The glowworms were so tiny that it took them several hours to eat up the small earthworm.

Glowworm was covered with slime from the earthworm. She pumped blood into her fringe foot until the bristly fringes stood out. With them she cleaned off the slime. As she twisted about, the two faint yellowish lights shone this way and that at the end of her body.

Glowworm could not bite through the hard coverings of ants or wireworms. Still, she found plenty of meat—more earthworms and cutworms and other animals with soft skins.

One night, some cows broke through a fence into the meadow. Wherever they walked, they packed down the loose soil. This crushed the tunnels of the tiny animals that lived there.

The heavy footfalls of the cows shook the ground. With each footfall, all the glowworms nearby shone their lights a little more brightly for a moment.

The next day the farmer fixed the fence, and the meadow was safe from the cows. If they had stayed, many of the soil animals would have died. They could not live in soil that was packed down tight. Nor could the grass get enough air and water in such soil.

But the cows did not stay. So the tiny animals kept moving through the soil. Their waste made it rich with plant food. The grass kept growing.

Glowworm grew, too, and shed her skin several times that summer. Each time, for a few hours afterward, her new skin was soft and

transparent. Then each time it hardened into a dark, horny shell around her soft body.

A year later, at the end of her second summer, Glowworm had grown nearly eight times as long as she had been when she hatched out of her egg. She was three-quarters of an inch long. When frost came, she went underground again and spent her second winter there.

Until spring, no enemy came near. Then a hungry skunk scratched deep into the soil, hunting for resting insects. He found many, but he didn't find Glowworm.

A SHELTER OF MUD

That night she came out of the soil and looked for food again. A few weeks later, when summer was about to begin, she did not look for food any longer.

Glowworm was now fully grown.

She found a spot of bare ground between two clumps of grass. There she stopped and dug out a mouthful of earth.

She chewed and mixed the earth for half a minute. Then she stuck her head out from beneath the shell of her body. She squeezed the chewed-up mud out of her mouth, onto the ground beside her. It came out like a short ribbon of toothpaste.

She dug another mouthful and chewed it. Then she squeezed out another ribbon beside the first.

Glowworm worked for three days and nights in that same place. She built the ribbons of mud up into a round wall, close beside her and all around her. Little by little, she built the walls up into a dome, making a curved roof over her head.

She took the mouthfuls of earth from underneath her body. Little by little, the ground became a curved floor.

Now she was in a round room with walls of dried mud nearly half an inch thick. Air came in between the pieces of mud, but the round room was dark and moist inside.

Glowworm became sluggish. In the next three days, her body swelled, and then her skin split down her back.

Out came a pupa (PEW-pa)—a strange, soft creature, looking something like a beetle. It was yellowish white.

CHANGING TO A FIREFLY

The pupa's wings were too tiny for her to fly, and her legs were too weak for her to crawl. But she did not try to crawl or fly. She lay on her back on the curved floor under the curved roof. Short stiff bristles in the pupa's back kept her from touching the damp ground.

The two pale glowworm lights were part of the pupa, and they still glowed faintly in the dark room. When she moved a little, the lights shone more brightly.

While the pupa lay there, she was changing into an insect that would no longer live in the soil. New long antennae and new wings and legs—strong ones—were growing under her skin. Two great round eye spots, each made of

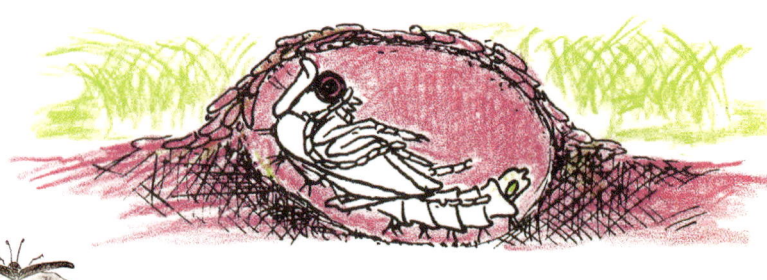

hundreds of tiny eyes, were growing outward from each side of her head. She would be able to see in nearly all directions.

Inside the pupa, eggs were forming and many new parts.

Breathing tubes grew from openings in the pupa's sides to all parts of its body. These tubes would bring in air containing oxygen.

Nerves grew, to carry messages and control movements all over the body.

Many nerves and breathing tubes ran into the light organs.

The light organs were growing on the underside of the abdomen, just beneath the skin. They were much larger than the old glowworm light. As the new lights grew, the old ones kept shining faintly. The old lights and the new ones were separate.

The light organs had light-making chemicals stored inside. When it came time for the light to flash, the nerves would carry signals into the organs. Then the chemicals would mix with each other and with oxygen from the air in the breathing tubes and with water from the firefly's body.

All these things working together in an instant would make the light.

Part of the light organ had tiny crystals in it. They reflected the light, helping it to shine more brightly. And the skin was transparent, to let the light shine through.

Already the new light was beginning to glow. But it did not flash out brightly yet.

After ten days all the changes in the pupa were finished. Its skin split open, and a beetle struggled out, just as the pupa had struggled out of the split skin of the glowworm. This looked like a real beetle, one that could crawl and fly—and make flashes of light, too.

It was a firefly.

DANCE OF THE SIGNAL LIGHTS

The next night, Firefly broke through the mud walls and crawled out among the grass blades. Her two tiny glowworm lights were still working, but they had begun to fade.

Many other fireflies were crawling in the grass. A few were flying overhead, flashing their greenish lights. But because the air was chilly and a heavy dew was forming, the fliers soon quit flashing.

Firefly spent that night under a fallen leaf. By the next morning, her old glowworm lights were gone.

At sunset, Firefly was still in her hiding place. The meadow was light, but the woods quickly grew dark. In this darkness of the woods, a few yellow firefly lights glowed.

It grew dark everywhere. The evening was still and warm.

On the edge of the meadow next to the

woods, more yellow lights flashed. These came from fireflies drifting slowly along just above the ground, dipping down and then up each time they flashed. Their flashes looked like the letter J written in moving light. As it grew even darker, more yellow lights came out. But Firefly stayed in her hiding place. It was not dark enough for her yet. The yellow lights were the signals of a smaller kind of firefly.

When it was nearly night over the meadow, most of the yellow flashers finished their evening flight.

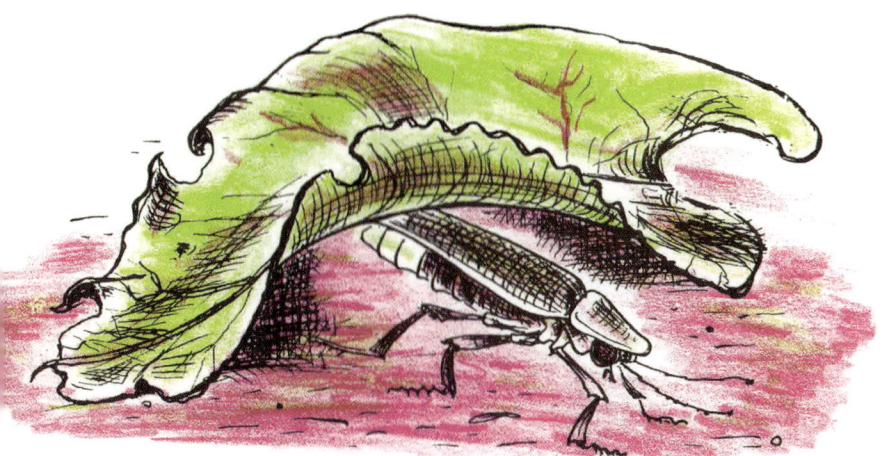

Now it was just beginning to be dark enough for Firefly. She crawled out of her hiding place in the grass roots.

A few feet in the air above her, a bright green firefly light flashed six times within half a second. This light was much brighter than the yellow flashes near the woods.

All over the meadow now, fliers were lighting up with their greenish flashes. They always gave five or six flashes quickly. Then after about five seconds, when they had flown fifteen or twenty feet, they flashed again. All these fliers were males of Firefly's kind.

More and more males flew over the meadow. The air was filled with their quick, bright signals.

Firefly crawled up on a tall grass blade. Other female fireflies were coming out of hiding places and crawling up stems and stalks and leaves as high as they could. They were going toward the dancing lights above.

A light breeze made Firefly's grass blade sway. She crawled to the tip and hung on in the darkness. The breeze blew against the flying males but not strongly enough to stop them.

Almost above Firefly, a flier gave a signal of six flashes. Firefly's large eyes saw much better than Glowworm's small eyes had. She caught the signal and answered with one short flash. Her nerves turned on the flash more quickly than electricity starts the light in a bulb.

Her flash lasted only a third of a second, but the male saw it. His eyes were even better than hers for seeing quick flashes of light. He turned at once in the air and dropped down toward her, flashing excitedly.

Firefly answered his signals.

Two more males flying nearby also saw her flash this time. Just as quickly, they flew down toward her, too.

The breeze caught all three males and blew

them past Firefly. They landed in the tall grass near her and kept flashing their quick signals.

Firefly answered most of their signals with her one flash. Each male watched for this answer and crawled through the waving grass toward it.

One of them crawled along a grass blade less than an inch away from her. But he did not know she was there because she did not flash at that moment. The fireflies could find one another only by seeing the light signals.

Finally, one of the males found Firefly. Now she had a mate. Neither of them flashed again that evening.

The other males that had been looking for Firefly saw no answering signal. Soon they flew away, to flash over the meadow and to watch for other lights in the grass below.

TRICKING THE YELLOW FLASHERS

Firefly's mate was a little smaller than she, partly because he had no eggs inside his body. He left her after a few hours, and she did not see him again.

Now that she had mated, Firefly's eggs would ripen, but she needed food. She spent the day quietly hiding on the underside of a leaf near the ground.

That night she went hunting.

Overhead, flying males were sending out

their signals. Firefly paid no attention because she had already mated.

She spread her stiff wing covers so that she could fly with the filmy wings tucked beneath. When she was in the air, she saw a few yellow lights of the fireflies that stayed near the woods. Some of the males were still signaling, even though it was dark and about time for that kind of firefly to stop flashing.

Firefly flew toward them and dropped down in the grass.

A late-flying yellow-flasher, only half as large as Firefly, came by slowly, signaling as he flew. Firefly answered with her own flash.

The strange male flew down toward her and signaled, and she answered again. When he reached her, she caught him in her jaws and ate him.

Firefly was still hungry. She flew again along the edge of the woods. A glowing light in a bush caught her eye. She alighted on the bush and crawled toward the glow.

It came from another male yellow-flasher caught in a spider web. The spider had bitten him, and the poison hurt him so much that he could not control his light. Little bright points of light gleamed from his light organ, here and there, so quickly that it looked lit up all the time.

Firefly tore the web. She was so big that the

spider did not try to stop her. She dragged the flashing male out and ate him, too.

She hunted for several nights.

Her eggs were ripening now, a few at a time. In the dark she walked over the meadow, stopping whenever an egg was ready. She pushed her long egg-laying tube into the ground and pushed the egg out through it and then walked on.

All through the meadow, hidden in the soil near the homes of the springtails and bristletails and the mites and the earthworms, her eggs were beginning to glow with faint yellow light. Soon they would hatch into new glowworms. In two years they would be fireflies, ready for their dance of lights for a few nights over the meadow.

THE END

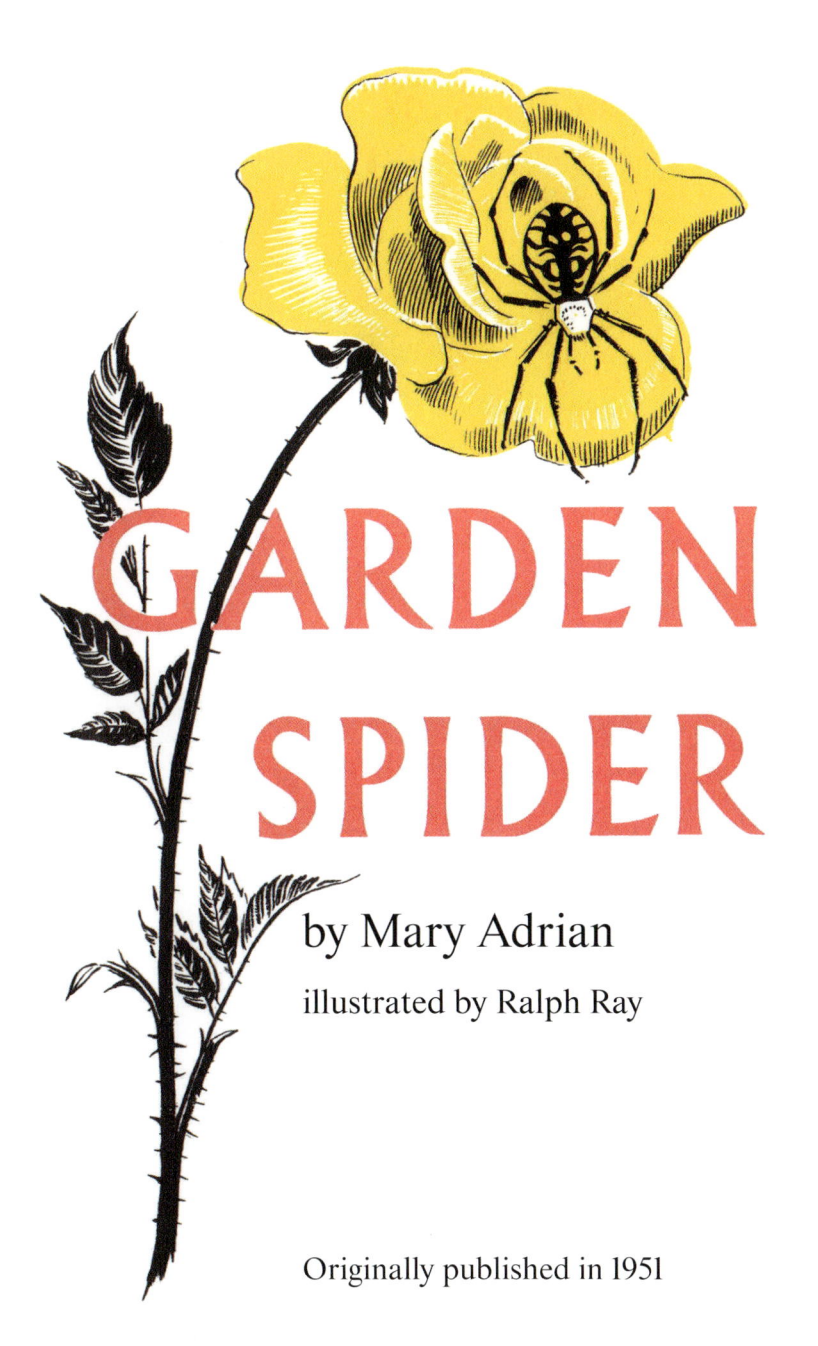

GARDEN SPIDER

by Mary Adrian

illustrated by Ralph Ray

Originally published in 1951

41

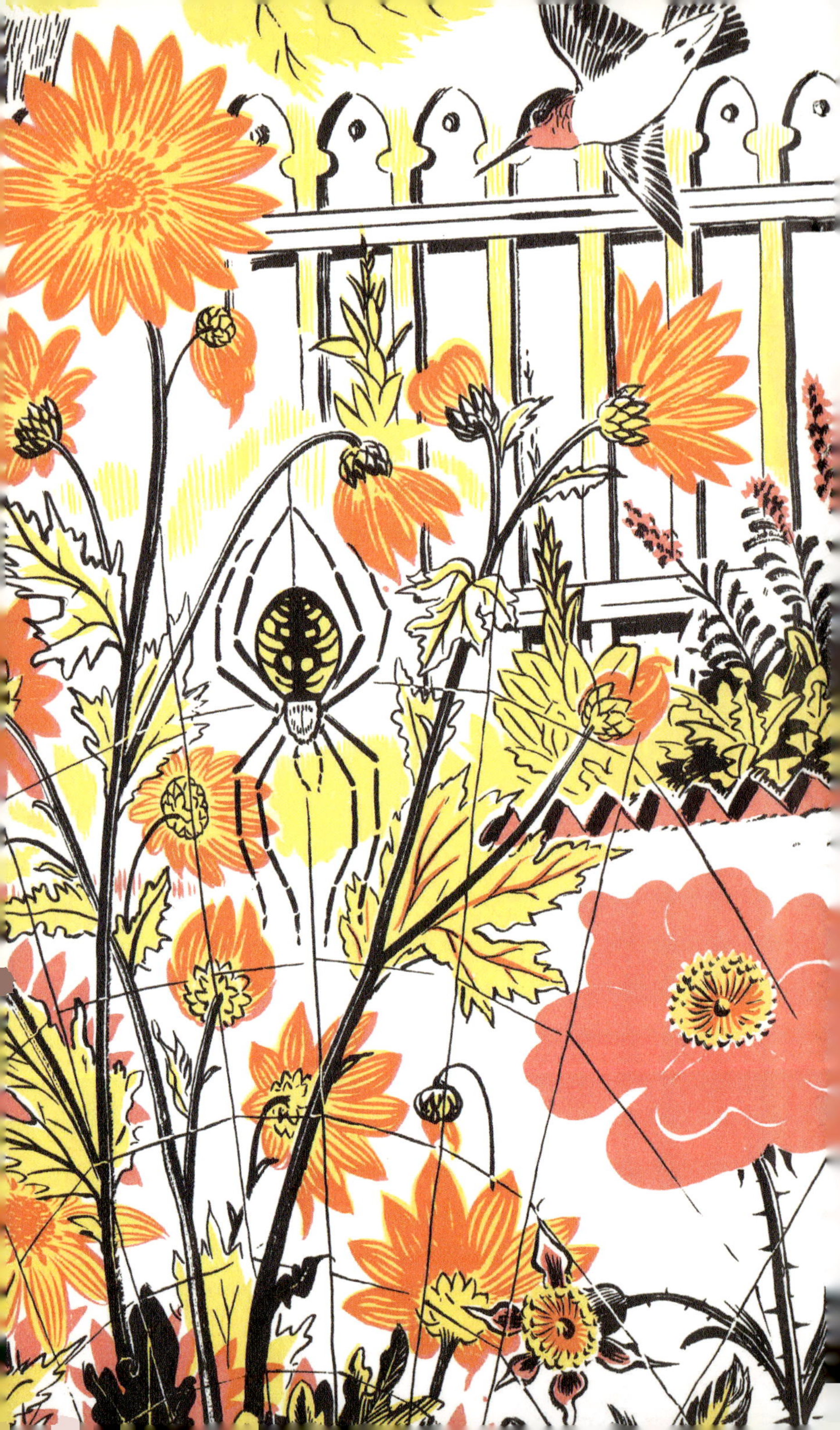

FOREWORD

Miranda aurantia, or the Garden Spider, is widely known. Its life history is typical of most spiders and has been given a great deal of study by our leading naturalists. It is to be found in gardens and fields throughout the eastern part of the United States, ranging from New England to the Dakotas, and south into the Gulf States. Close relatives are to be found in other parts of the United States.

I wish to express my thanks to Mr. John C. Pallister, Research Associate, of the American Museum of Natural History for checking the manuscript for scientific accuracy.

Mary Adrian
Darien, Connecticut

44

FIVE HUNDRED EGGS

A garden spider was climbing up a dying flower stalk. Many of the plants in the garden around her were turning brown.

Garden Spider was feeling the frosty fall air against her body. This frosty air was a signal to her that the great time had come. It was time to lay her eggs.

She kept climbing upward. Her eight legs easily carried her oval-shaped body.

When the eight legs stepped upon broken stems or dried leaves, Garden Spider did not feel the jar, even though her body was heavy with the eggs. That was because her springy legs held her body snug between them.

She had a beautiful body, this creature of the garden. It was black with gold spots. Very, very fine hairs grew all over it and down her legs. Like most spiders, she had eight bright eyes, set in two rows on the front of her head.

Even with eight eyes Garden Spider could not see very far. But the fine hairs on her legs and body were almost like far-seeing eyes. They could feel the slightest movement of everything nearby, even of the air.

Up the flower stalk she climbed, stepping carefully and feeling her way. Then she stopped to spin her egg case.

Garden spiders are spinners and wonderful ones. They spin silk, and it comes from their own bodies. It comes out of spinnerets, which are tiny tubes under the back part of the body. While the silk is inside, it is a liquid, but as soon as it comes out, the air hardens it into a delicate thread.

First, Garden Spider spun several crosslines of silk and attached them to different parts of the flower stalk. These made the frame of her nest.

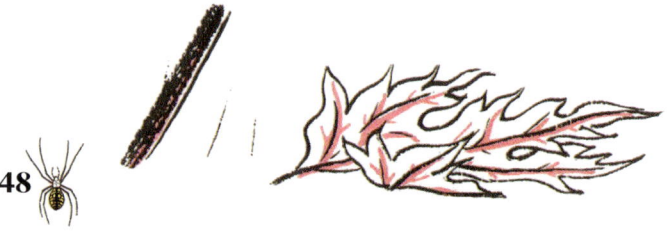

She felt the leaves tremble in the cold wind, but she did not stop. She spun a roof of silk threads above her. It looked like a bell.

Now her great time had come.

Garden Spider hung downward and forced a sticky liquid from her body up against the tiny silk roof. Then, while she was still in this position, she laid her mass of eggs in the sticky liquid. They held together in a tiny ball that hung down from the roof.

Garden Spider must have been proud. She had a right to be, for she had laid five hundred eggs in about ten minutes.

Then she spun six blankets of silk around her eggs and the roof. This made a snug nest. It was shaped like a small pear. It was even light yellow like a pear.

Making this egg case took Garden Spider several hours, even though she made only one. Now she was very tired, and her body was thin and shriveled. Like all garden spiders, she became old after she had laid her eggs and spun her nest. But she clung to the flower stalk to guard her eggs.

In a few days, the life in Garden Spider's body came to an end. Her body fell to the ground.

PEARLS WITH LEGS

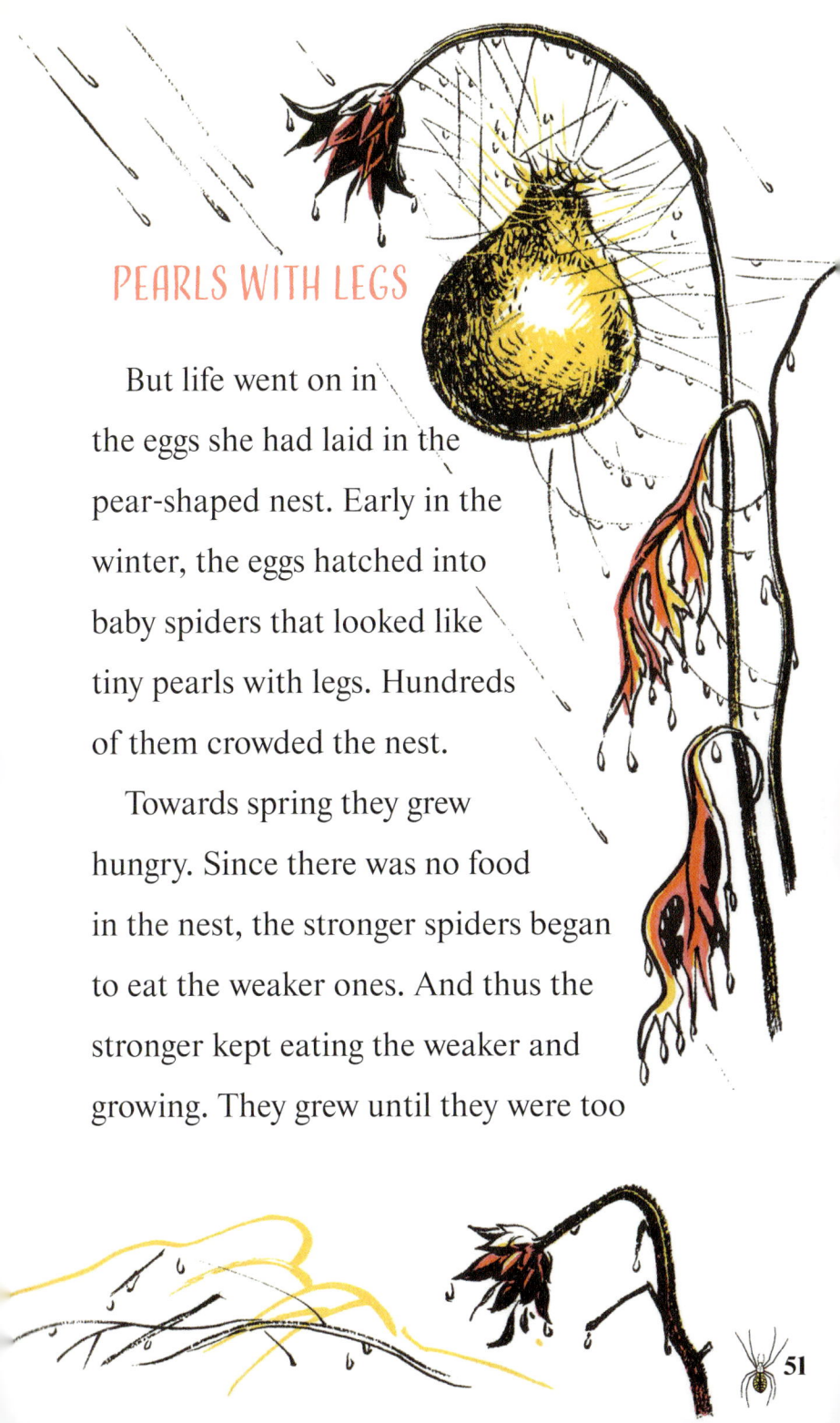

But life went on in the eggs she had laid in the pear-shaped nest. Early in the winter, the eggs hatched into baby spiders that looked like tiny pearls with legs. Hundreds of them crowded the nest.

Towards spring they grew hungry. Since there was no food in the nest, the stronger spiders began to eat the weaker ones. And thus the stronger kept eating the weaker and growing. They grew until they were too big for their skins and had to shed them. They grew darker in color.

When spring came, many of the little spiders had been eaten, but there were still dozens of them left. They felt an urge to leave their nest.

One spider cut a hole in the roof. She squeezed her tiny body through the opening. The others followed. Soon the flower stalk was filled with tiny spiders.

They crawled down to the ground, their dark bodies touching many green plants and budding flowers.

The little spiders began to explore. Although each went off alone, they did not get lost from one another. Wherever they went, they spun out silk lines from the starting point. When dusk came, each little spider followed its line back to the others. Then they huddled together for the night. They felt strange away from the nest.

IN THEIR OWN PARACHUTES

In a few days, however, the little spiders knew they were ready to separate and go out into the world alone. And they also knew just how to do it.

It was a great day for them. Each tiny spider crawled up a tall blade of grass and spun out several lines of silk from its spinneret. The ends of the silk lines floated out into the air.

A strong breeze tugged and pulled at the lines. The spiders let go of the grass and held onto their silk lines. The breeze lifted them, and they sailed into the air like tiny parachutes.

Higher and higher they sailed until they were above the treetops. Now the breeze carried them over the fields and brooks and roads and blew them apart from each other.

One garden spider saw a swallow fly by, but the swallow did not see her. If he had, he would have eaten her. But she was only a speck in the sky, and her silk lines could hardly be seen in the sunlight.

The breeze blew the little
spider over a town. Swiftly she
floated by buildings, some low,
some tall. She almost hit a church steeple.
Then the breeze began to die, and she began to
float gently downward. She landed in a flower
garden.

AROUND AND AROUND AND ZIGZAG

Young Garden Spider was very hungry after her journey. Since there were no smaller spiders around to eat, she had to get food in some other way. She knew just what way that was. She would have to make a web that could trap insects.

She climbed up a rosebush. Her mother was not there to show her how to build her first web. Yet she knew how.

First, Garden Spider spun out a silk thread that floated into the air and caught upon a twig of the rosebush. It made a bridge. She ran across it, adding other threads to strengthen it. Then she spun out more threads that reached

57

to other twigs, to make the frame of her web.

But what was she going to put inside this roundish frame? Little spokes, like the spokes in a bicycle wheel.

Garden Spider spun a line across the open space of her frame. She walked along the line to its center. There she attached a new line, and then she returned to the outer edge of her web, spinning the line longer as she went. She was making her spikes now—silk lines running from the center to the outer frame.

Every time she made one of the spokes, she walked on another spoke that she had already made. Every time she went through the center of the spokes, she added threads that tied them all together.

By the time she had finished making the spokes and tying them together, she had a cushion in the center. The cushion was big enough for her to sit on.

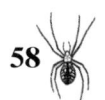

Next, beginning near the center, she spun a spiral thread round and round from spoke to spoke. When she had filled in all the spaces, she rested a few minutes.

There was more work to do before she could eat.

The web she had made so far was not a web that could catch insects. An insect might fly against the web, but it could also fly away again.

Garden Spider knew that a sticky web could hold the insects fast. She knew that she had sticky liquid in her body. She could spin plain silk, or she could spin sticky silk. So now she began to spin a whole new set of spiral lines—sticky ones.

This time she began at the outside of her web. She spun her sticky spiral thread round and round from spoke to spoke.

As she spun, she walked on the old spiral

lines so as not to get herself caught on the sticky lines. As she made the sticky spiral lines, she cut the old ones away. She had made the plain ones just to walk on, and now they were of no use.

Then she decorated her web. She spun a zigzag ribbon of silk across the center.

Her web was finished. It was only two inches

across, but it had fifteen spokes in it. It was a beautiful web, one to be proud of.

Garden Spider was glad to have it done. She was very, very hungry by this time, and now she would be able to catch food.

She set about oiling herself so that she could run over the sticky lines without getting caught. It was not easy to put oil on herself. She had to draw her eight legs, one at a time, through her mouth, where the oil was. Then she rubbed oil on her spinnerets and parts of her mouth since they also might get caught.

Dusk was now settling over the garden. Garden Spider lay on the cushion in the center of her web and waited. It was not long before the lines began to shake. A mosquito had struck the web. He kicked, trying to free himself from the strong sticky lines.

Garden Spider ran quickly across her web. She pounced upon the mosquito and bit him.

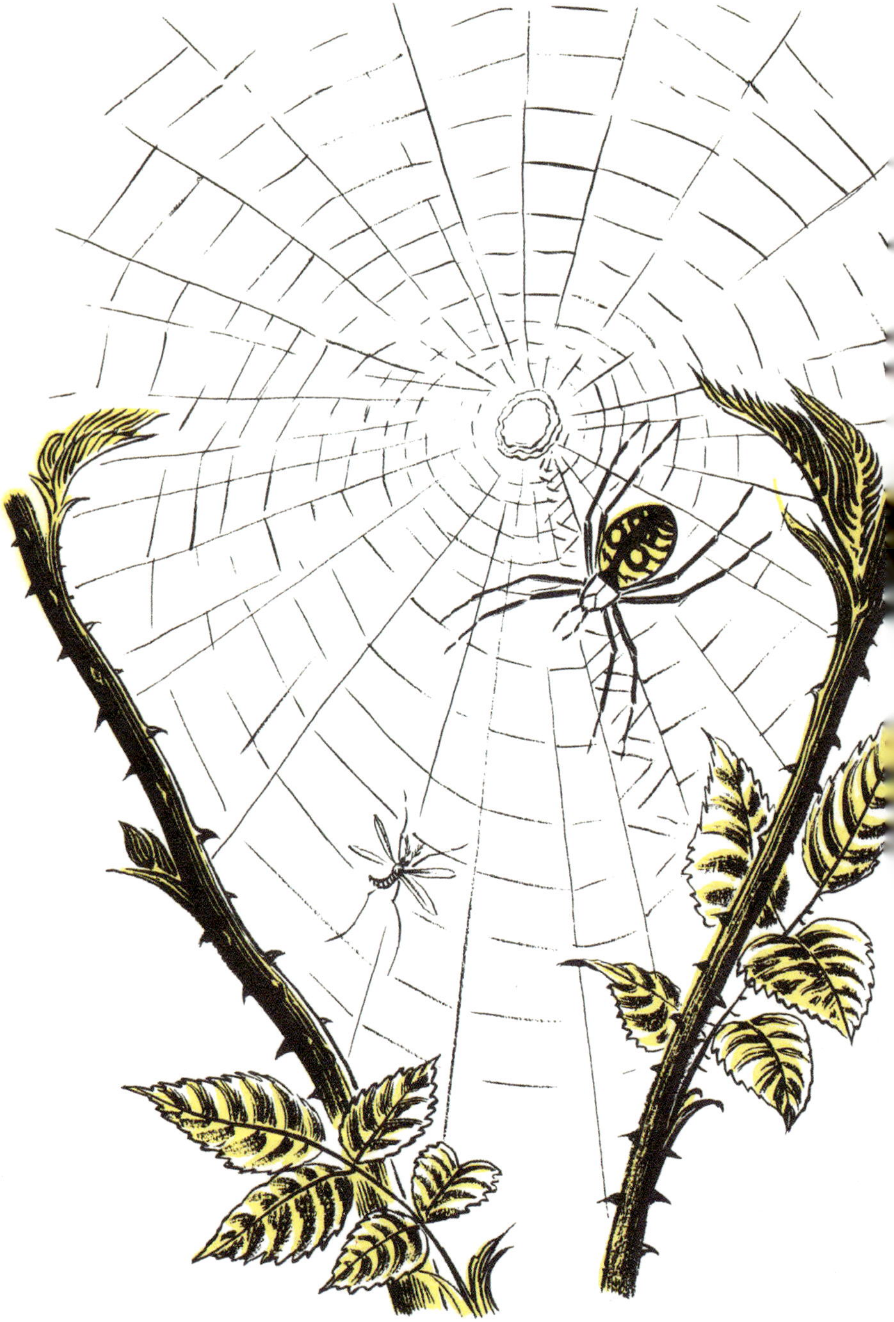

From her fangs, tiny drops of liquid went into the mosquito. He died. This liquid is poison to insects, but not to animals or people.

Garden Spider then crushed the mosquito with her pedipalps, two tiny feeler legs on either side of her mouth. After that, she sucked the body fluids. She could take only liquid food, for her mouth was too small for any other kind.

Then she dropped the mosquito out of her web. Several strands of silk had been broken. She mended them neatly and went back to her cushion to wait for another insect. She was still hungry.

NEW SKIN FOR OLD

Days went by. Garden Spider felt the sun grow warmer. She had grown since she had parachuted down into the garden, and so her black and yellow skin was too small for her. She was going to shed it, as she had done in the nest when she was very little.

She felt her skin split at the sides and near her eyes, too. She worked her legs out of the old covering, one by one. In a few minutes, it lay alongside of her, an empty shell.

Garden Spider's new skin was pale and very soft. She was weak and could not move fast. Her enemies could easily attack her, and so she rested and waited for her skin to harden. It took several hours, but nothing disturbed her. How relieved she was! And how good it was to have room inside her skin.

THE WEB THIEF

One morning Garden Spider felt a sudden pull at her web. She felt that the pull was too strong for an insect.

She was right. A hummingbird was pulling at the web with her needle-like bill. She wanted the silk for her nest.

Garden Spider was afraid Hummingbird might wish to eat her. Hummingbirds do not eat garden spiders, as do wrens, but Garden Spider didn't know this. She did not take any chances.

She ran quickly to the other end of her web. She spun out a dragline and attached it to the rosebush. Clinging to the line, she dropped. Down, down she fell. All the while she kept spinning more of the dragline to break her fall.

The rosebush shook as she floated to the ground, for Hummingbird had begun to tear the web apart. Garden Spider did not care about her web now! All she cared about was hiding! When she reached a plant near the ground, she scurried under a leaf.

Hummingbird flew back to her nest with the silk of Garden Spider's beautiful web in her bill. Garden Spider peeked out of her hiding place. Everything was clear, so she started to come out. Then quickly she ran back.

67

Flying overhead was a spider wasp. She was looking for a spider to feed to her young. Her deep purple wings were moving swiftly.

Garden Spider held ever so still. Not even the fine hairs on her legs and body moved. After a while, she peeked out of her hiding place again. Spider Wasp had flown away.

Garden Spider quickly took hold of her dragline that still hung from the rosebush. Up she climbed. She had climbed only halfway up when again the wasp's purple wings zoomed overhead.

Like a streak of lightning, Garden Spider dropped to the ground on her dragline. She thought that the wasp had seen her. But instead, Spider Wasp saw a grass spider.

Spider Wasp stung the grass spider. It could no longer move. Then she dragged it to her nest in the garden wall, where she kept other paralyzed spiders.

Little did Garden Spider know that Spider Wasp was going to lay an egg on the paralyzed grass spider, and that when the egg hatched, the baby wasp would eat the spider.

But Garden Spider did know that she would have to be more than ever on the alert for enemies. Again she climbed up the dragline.

She found a new place for a web in a flowering bush. Here she spun a new web. Because she was now a grown spider, the web was very large. It was two feet across and had twenty-eight spokes in it.

THE MATE

More weeks went by. It was late summer. Garden Spider had shed her skin many times. She was full-grown now, an inch long. The yellow spots on her black body glistened like gold in the sunlight.

At dusk one evening, someone came to call on her. A male garden spider tugged at her web. He looked like a dwarf compared to her, for he was only a quarter of an inch long.

Slowly the male spider walked over the strands of silk in the web. Garden Spider watched him for a moment. Then she pulled at the silk lines with her legs. It was a warning to him not to come any closer.

The male spider backed away. In a few minutes, he rushed forward. This time Garden Spider did not give another warning. She

71

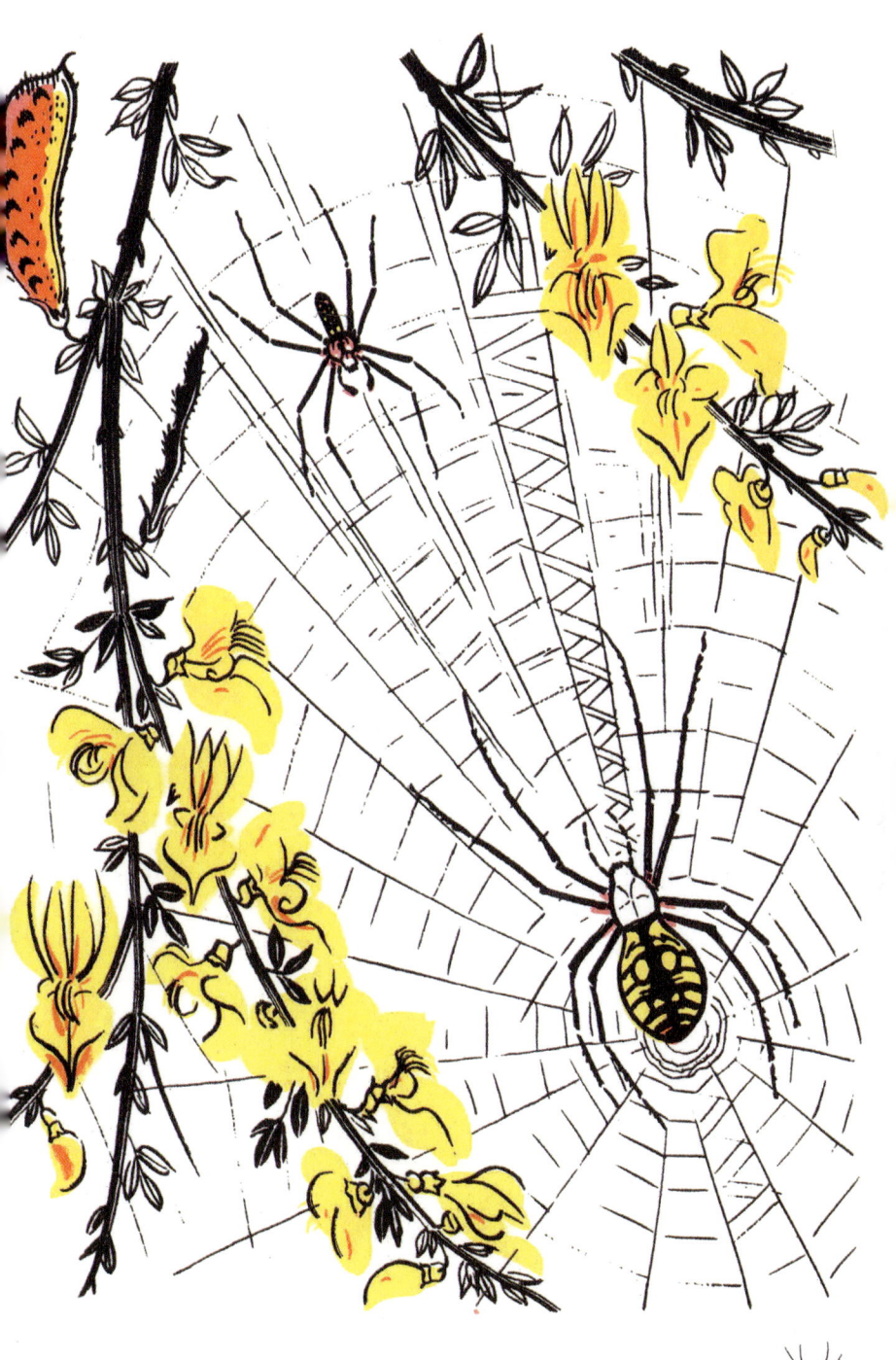

pounced upon the male and ate him, for she was very hungry.

The next evening another male spider tugged at her web. She did not warn him not to come. Step by step he came closer. He was ready to drop from the web, though, at the slightest warning from her. But Garden Spider was not hungry this evening. She let her visitor stay, and they became mates.

After a few days, the male spider went away to die, like all male spiders that have been mates. If he had not gone away, Garden Spider might have eaten him.

BUMBLING BUMBLEBEE

One day Garden Spider lay on her cushion in the sun, half asleep, when she felt something strike her web.

Instantly she was awake and on guard. A huge bumblebee had hit her web. He was kicking furiously to free himself.

Garden Spider knew that this was no easy insect to capture. Bumblebee was larger and stronger than she, and he had a powerful dagger.

In great excitement, she grabbed hold of the web and shook it. Bumblebee kept kicking his legs and moving his wings. His struggles and her shaking of the web imprisoned him even more.

She ran toward the huge body, careful to keep away from the powerful dagger. Her spinnerets began to spin out a broad band of silk. She threw it over Bumblebee. She threw another and another and another. Then she grabbed him by the tips of the wings.

Bumblebee struggled for his life. He rocked the web back and forth, but Garden Spider held his wings tightly. Then, quick as a flash, she darted to the back of his neck. In went her sharp fangs, sending the tiny drops of poison.

A few seconds later, Bumblebee was dead. His legs hung lifeless. His tongue stuck out.

Garden Spider was still excited. She had used all her skill to capture this large insect. Now she would have food for several days.

FIVE HUNDRED EGGS AGAIN

More weeks went by. The nights grew cooler, the days shorter. Garden Spider was ready to lay her eggs.

She climbed up a dead flower stalk just as her mother had done the fall before. Up the flower stalk she climbed, stepping carefully on her eight springy legs. Near the top, she stopped to spin her nest.

She laid her five hundred eggs. She spun six blankets of silk around the precious eggs and made a snug nest. It was shaped like a pear.

In a few days, the life in Garden Spider's body came to an end.

Her body fell to the ground.

But Garden Spider's life went on in the pear-shaped nest. The two lives of herself and her mate had become divided into the hundreds of lives in the eggs.

One day next spring, tiny garden spiders will leave the nest and spin their webs and lay their eggs. It will be the same story as this all over again.

THE END

HONEY-BEE

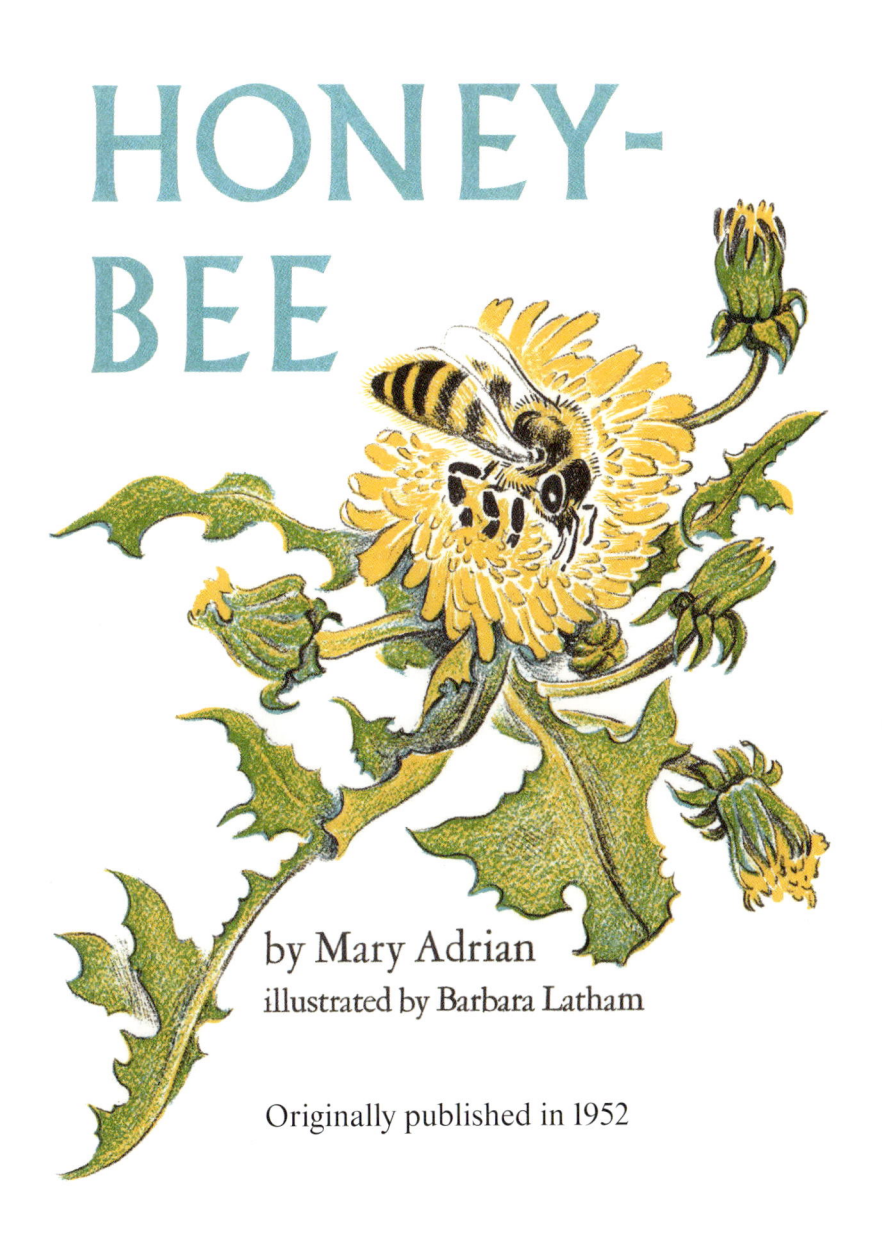

by Mary Adrian
illustrated by Barbara Latham

Originally published in 1952

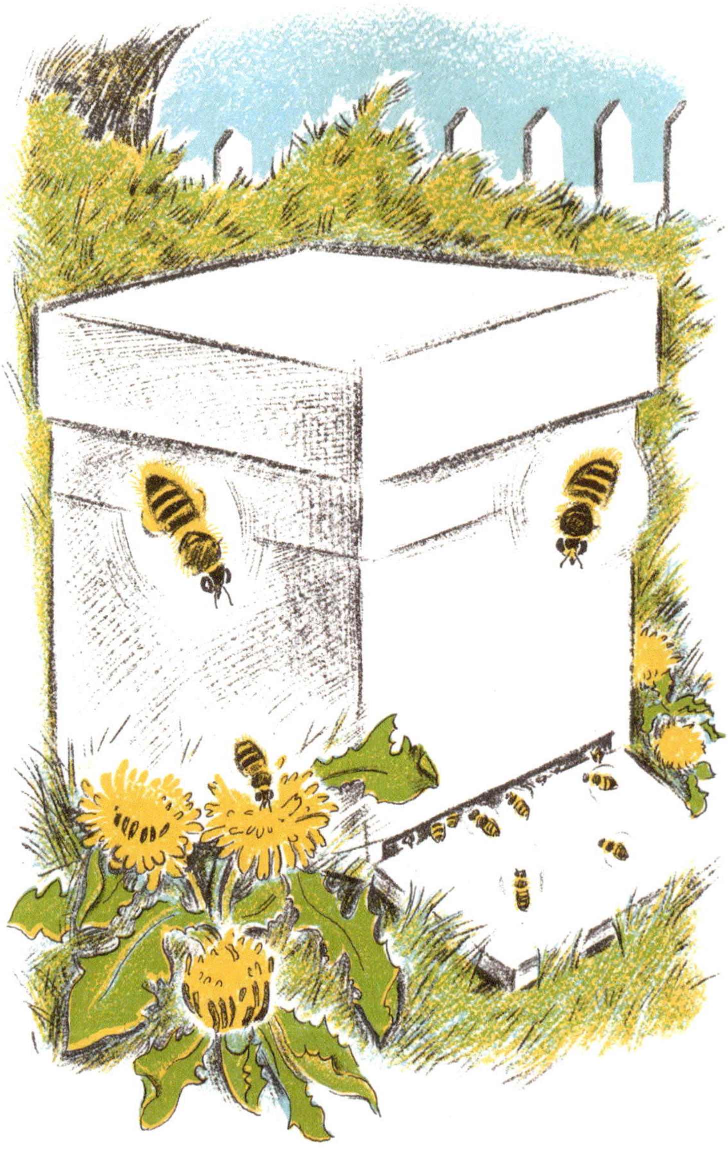

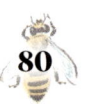

FOREWORD

Of the thousands of different species of bees in the world, the honeybee, *Apis mellifera*, is the most widely known. There are several races of honeybees differing in size, disposition, and color. The gentle yellow-banded Italian honeybee is the most common in the United States. It was imported into this country many years ago.

I wish to take this opportunity to thank Dr. Donald R. Griffin of the Department of Zoology, Cornell University, for his advice and helpful criticism.

Mary Adrian
Darien, Connecticut

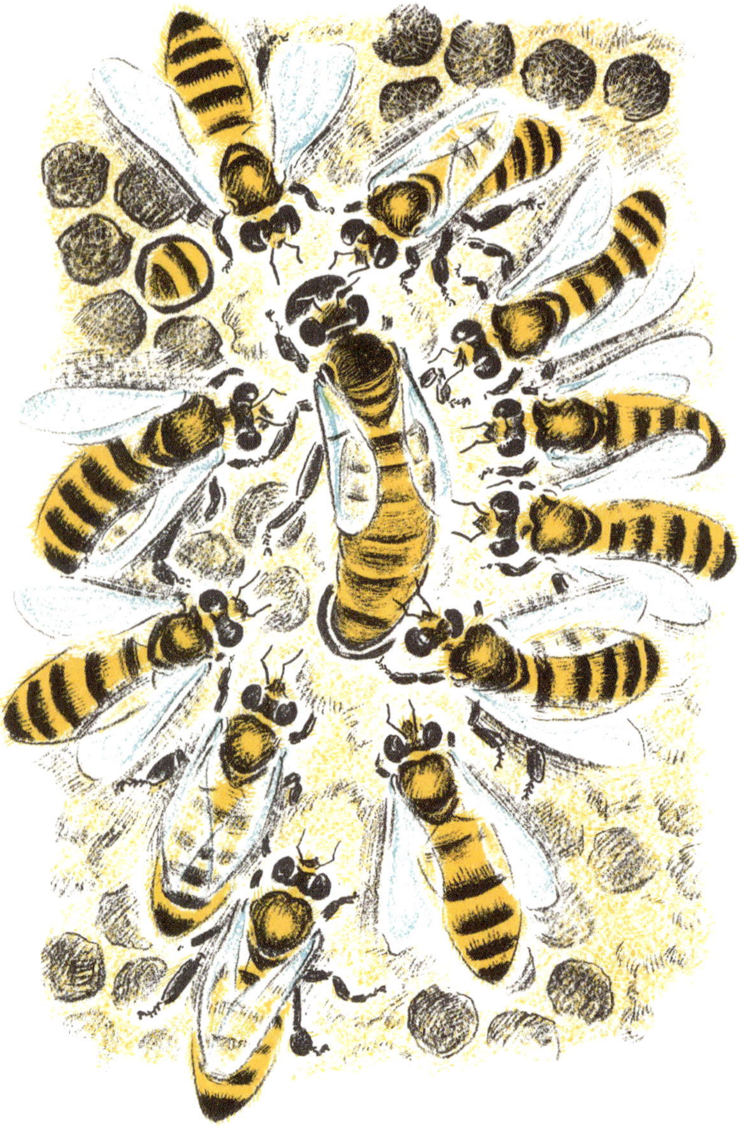

A QUEEN LAYS AN EGG

Humm-Humm! Honeybees were working in their hive one spring day. Thousands and thousands of bees were crawling over the combs, working and humming. It was dark in the hive, but they smelled and felt their way around with their long antennae.

Wax combs, made of tiny cells, hung from the ceiling in wooden frames.

The queen was on one of the combs, laying eggs. Her long, slender body was banded with gold. She was a handsome insect, as the queen and mother of 6,000 bees should be. Small worker bees kept circling around her and touching her with their antennae.

The queen put her head into a cell to see if it was clean. Then she turned around and laid in it a tiny bluish-white egg.

For three days the egg lay in the cell. Then it

moved, and out crawled a white larva with a black head.

The larva looked like a worm. She had no legs. She had no eyes. But she had a mouth, and she was hungry.

So she ate a few drops of bee milk from the forehead glands of a nurse bee who came to feed her.

The larva liked the sour taste of the bee milk, but the nurse bees fed it to her for only two days. If she had kept on eating bee milk (or royal jelly, as it is also called), she would have grown into a queen bee. The hive needed workers, not queens. So on the third day, the nurses began to feed her bee bread, a mixture of pollen and honey.

The larva liked this, too. She ate so much of it that she grew too big for her skin. *Pop!* Off it came. She ate more and more and more bee bread and shed her skin nearly every day.

On some days the nurse bees fed her a thousand times. In five days she had grown so big that she filled the cell.

The larva spun a cocoon. Round and round she covered her round body with white silk from her mouth. Then she lay very still and went to sleep.

Wonderful things began to happen to the larva. Her round body began to separate into a thorax (or a chest) and an abdomen. From her head grew two antennae and five eyes—two great ones on the sides and three tiny ones on the forehead.

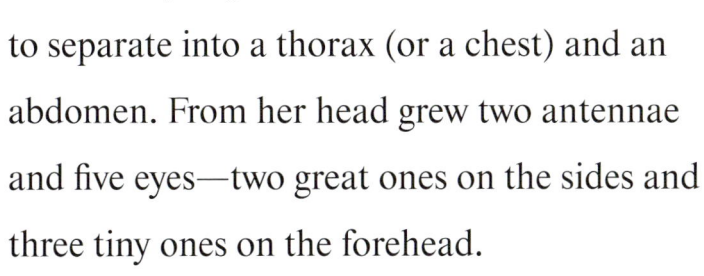

The food she had eaten was being used for all this growing. She was still white, but she began to look like a bee—a bee carved in wax.

From her thorax grew three pairs of legs and two pairs of wings. On the outside of her body a soft shell began to form.

Last came her coloring. Her great eyes grew pink then black. Her body became banded in gold and black and covered with fluffy yellow hairs.

This little insect did not know that she had slept twelve days and changed from a larva into a bee. But she did know that she wanted to leave the cell.

She kicked off her cocoon and bit a hole in the wax cover of the cell.

She poked out her head with its pretty oval

face, great black eyes, and long antennae. It was too dark in the hive for her to see anything, but she smelled many rich odors.

She crawled out of the cell, onto the comb of the busy, humming hive.

AWAKE AND HUNGRY

All around Honeybee hundreds and hundreds of other workers had just awakened, too. They were a little weak in their soft shells, but they began to lick their hair smooth and to clean their antennae.

Older bees, rushing by, ran against them. Honeybee held fast to the comb with her claws and sticky footpads to keep from falling off, down to the bottom of the hive.

She crawled into an empty cell. There she stayed while her shell hardened into a kind of outside skeleton.

Hours later, she crawled out of the cell and over the comb.

All around her thousands and thousands of bees were working and humming.

Honeybee kept waving her antennae, smelling the many odors of the hive and feeling her way. Each bee she met had a different odor.

She came to a richer smell and stopped.

It was the queen close by, laying eggs. Today she was laying in big cells, for these eggs would hatch into drone bees. Drones are males and much bigger than workers, who are females.

Honeybee poked her head into an empty cell and began to polish it. Soon she had it clean enough for the queen to lay an egg in.

The queen stopped to rest. The workers began to feed her.

Honeybee had not eaten for twelve days, and she was hungry, too. A delicious new smell came to her and made her even hungrier.

She walked in the direction of the new smell, off the nursery comb, and onto the food comb. There she found honey. So that was what had the delicious smell! From an open cell, she ate her fill.

Here, too, all around her, thousands and thousands of bees were working.

Some were storing honey in the cells. Some were sealing the full cells with wax. Some were bringing in honey and pollen from the fields.

Honeybee poked her head into a cell and found lumps of pollen in the bottom. She smoothed them down. She liked the smell of the pollen and tasted some. It was as delicious as honey.

Now that she knew where the food combs were, she ate whenever she was hungry.

NURSE BEE

In a few days, the glands in Honeybee's forehead were full of bee milk. So she fed the larvae with her milk and became a nurse bee.

She also fed them bee bread, running back and forth from the nursery combs to the food combs. She fed hundreds and hundreds of hungry larvae and made thousands of trips. As she worked, she beat her wings, which made a soft humming.

She got very tired and sometimes rested a little. Often she stopped to clean her pretty face and to brush her soft yellow hair with her front legs.

Soon the drone larvae hatched from the male eggs the queen had laid. Honeybee and the other nurse bees fed them, too, the same food as the worker larvae.

PLAY AND DANGER

Honeybee was over a week old before she left the dark hive. Waving her antennae to feel her way among the thousands of other bees, she crawled down to the floor.

She went out the opening and found herself on a landing board among other workers.

Some were guarding the entrance. Some, like herself, were going out for the first time. Some were landing on the board as they flew in from the fields. Some were cleaning the ground nearby, carrying away dead bees.

Honeybee liked the warm sun. She liked using her eyes. She could see the grass and the trees and the sky, all at once.

She turned around and looked at her hive. She saw one side of a white box with a long low entrance. She walked back and forth, looking at the hive.

Then she sprang into the air a few inches and flew from side to side, still looking at the hive. She flew higher. Now she could see all of the white box, and green grass and yellow dandelions around it.

Then she circled in her flight. Near the hive she saw a fence and an apple tree full of white blossoms. She must keep a picture of them in her mind so that she could find her hive.

Honeybee liked to push her lacy wings against the air. The minute she had sprung into the air, she had known how to fly. Her back wings had hooked themselves to her front wings, making one big wing on each side. She beat them so fast that they made a loud buzz.

She kept circling—not far, but where she could see the apple tree or the fence or the hive. Flying was not work. It was play. She buzzed happily.

All around her thousands and thousands of other young bees were flying and buzzing.

She darted among them. They flew up in the blue sky toward the sun. They flew down toward the green grass. They swirled in and out and around each other. They flew backwards. They stopped in the air and hovered, like tiny helicopters. They sailed, and they zig-zagged.

Some drones were out flying, too. Honeybee saw one pass her and go zooming up, higher than the apple tree. She would never be able to fly so fast, no matter how much she practiced. Drones were big and husky, and almost as long as the queen.

Suddenly a dragonfly dropped from the sky into the dancing bees. He scooped up one of them in his six legs, which he held together like a basket. Off he flew, eating the bee.

Honeybee and the others tried to crowd inside the hive, but the guards held them back until they had smelled each one.

Guards had to watch for strange bees, who might come to steal honey from the hive.

They also had to watch for other robbers: mice and men and skunks. They had only one weapon, their sting. When they used it, it took their lives—for a worker cannot pull her sting out of tough skin. She tries to, but the pulling tears the sting out of her body, and she dies.

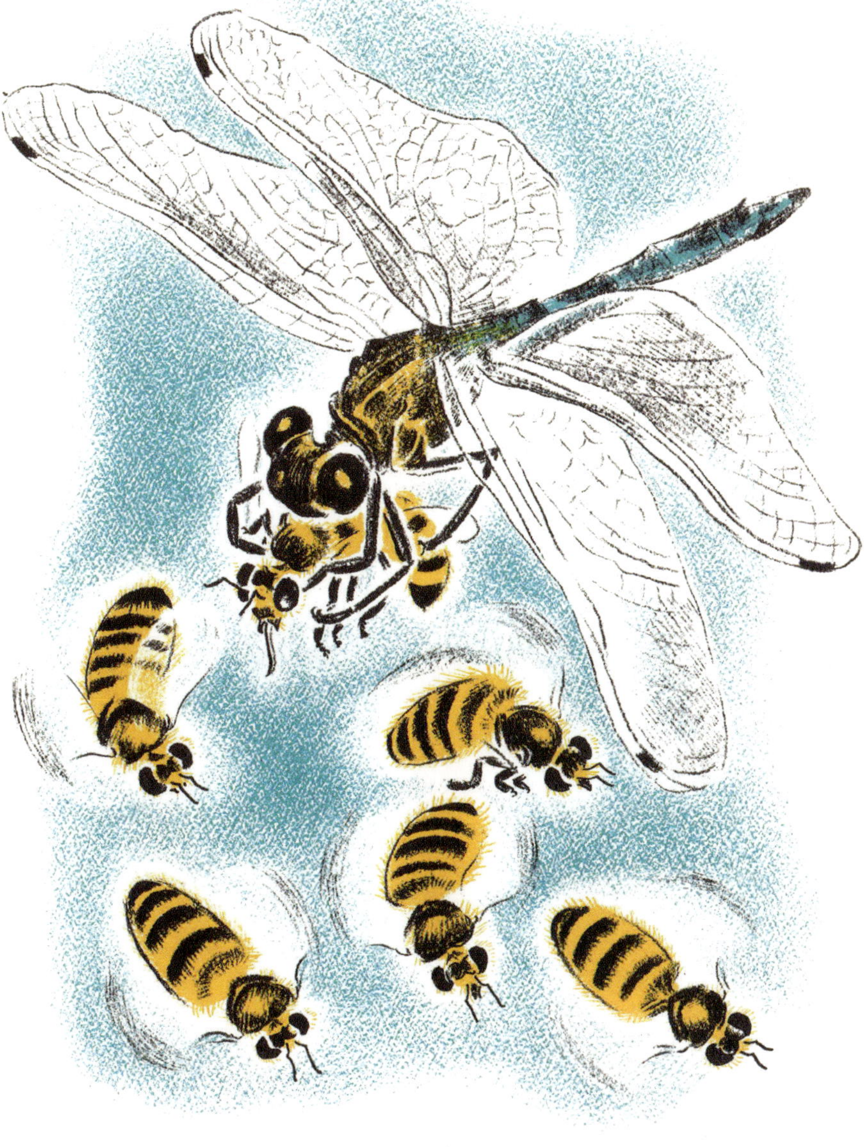

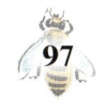

MAKING HONEY

As Honeybee grew older, she had less and less bee milk. She did more of the other work in the hive, and so she became a house bee.

One day a field bee came in with a load of honey. She carried it inside her body in her honey sac.

She came up to Honeybee on the comb and stuck out her tongue. Honeybee was not hungry, but she opened her mouth anyway. Field bees sometimes fed her nectar. But this one was giving her new work to do.

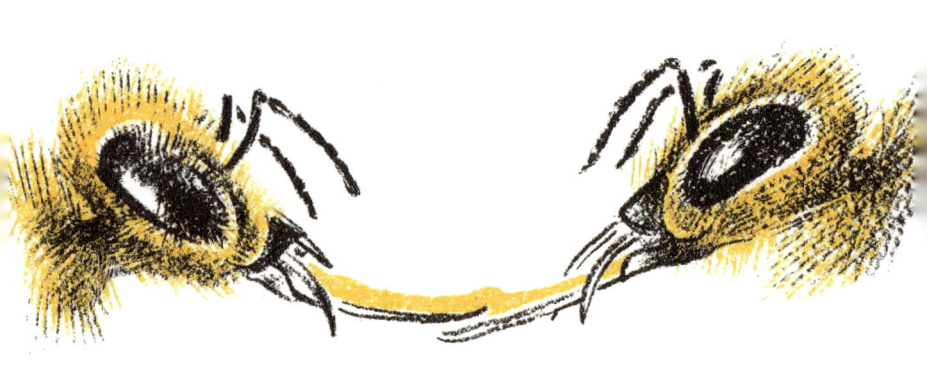

It was the work of making honey. The nectar had begun to change into honey while the field bee was carrying it home in her sac. She had to go back to the fields, and so she gave the nectar to Honeybee to finish changing it into honey. It flowed into the young bee's mouth and down into her honey sac.

The field bee left to gather more nectar. Honeybee stood quietly. She uncurled her long, tube-like tongue and rolled a droplet of nectar out along it. She curled her tongue up again, and the droplet rolled back into her honey sac. She did this over and over until all the nectar had been rolled out and back, and made into honey.

Then she put it in a cell and went outside to practice flying.

But the honey-making went on. Thousands and thousands of field bees were bringing in nectar and giving it to thousands and thousands of house bees. Soon all the empty cells of the food comb would be filled.

The queen was laying nearly 2,000 eggs every day in the cells of the nursery comb. Soon the cells would all be filled.

The hive needed more cells.

MAKING WAX

Several hundred workers went to the ceiling and hung themselves there in a wide chain. Some clung to the ceiling. Some clung to the bodies of other bees. These workers were going to make wax for new cells.

When Honeybee came in from flying, she had to help because she was old enough for her wax glands to work. But first, her glands needed plenty of honey for making wax. So she ate and ate all she could hold.

Then she crawled to the wide chain of bees and wound her legs around one of them. She pressed her body very close.

Hours went by. Night came.

Honeybee and the others felt very hot. Not much air could get to their breathing tubes. They got hotter and hotter. More hours went by. It was almost morning before they were hot enough for the wax to come.

Honeybee felt the wax ooze from the glands on her abdomen and run into pockets on her body. It hardened into scales.

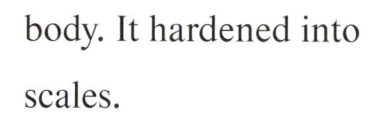

She broke away from the chain of bees. With the hairs on her hind legs, she lifted the scales out of her pockets, passed them to her front legs, and into her mouth. Then she chewed them and mixed them with saliva until they became creamy white wax.

Honeybee put the wax under her chin to keep it warm and went down to the food comb. She plastered the warm wax against one side of the comb, where more honey cells were needed. She smoothed it on one side while another worker smoothed it on the other. Then she ran off.

Another wax-maker came along, gave the

soft wax a pat, and ran off. Then came another, who pinched the wax a little. Then another, who added her bit of soft warm wax. Each wax-maker gave a pat or a pinch or added a bit more wax, but none of them stayed to do more. In this way, thousands of bees shaped and smoothed the wax into cells as thin as paper, and with six sides.

Honeybee had run off to rest. Wax-making was even harder work than feeding larvae.

She found a place on the frame near an old bald-headed worker, who was asleep. Honeybee, too, went to sleep. Like the eyes of the old bee, her eyes were wide open. She had no way of closing them.

FASTER THAN ELECTRIC FANS

Honeybee woke up and felt about with her antennae. The old worker was not there. She had died in her sleep and fallen down to the floor of the hive. She was six weeks old and had worked too hard to live any longer.

Honeybee, too, would be an old bee with a bald head at five or six weeks of age. Then she, too, would die. But now she was full of life. She brushed her yellow hair and went back on the honeycomb, ready to work again.

A hungry drone came to her. Honeybee fed him. She seemed glad to do this and a little proud of the handsome fellow. Perhaps she was sorry for him. His tongue was too short to gather nectar from flowers. He had no wax-making glands. He had no sting. The drone ate his fill and then went outside to fly.

The air in the hive was hot. It was too hot for the larvae. It was too hot for the honey to ripen in the open cells.

Honeybee went down to the entrance, where she found several young workers.

They all gripped their claws into the floor. Then they beat their wings very fast—faster, faster, faster—faster than electric fans. They fanned the hot air out of the hive and cool air into the hive.

Fanning was hard work, but their wings were strong from the practice-flying.

Older workers brought in drops of water and placed them on the frames.

The hive became cooler and cooler. The bees stopped fanning.

107

ROUND DANCE

When Honeybee was two weeks old, she was a grown bee and ready to gather pollen. But she did not know where to find it.

The field bees knew. One of them had just come in with pollen and was telling the other workers where to find it. Honeybee went up to her.

The field bee's way of telling was to dance. She was doing a round dance. She turned all the way around to the right, making a circle. Then she turned back, on the same spot, all the way around to the left, making another circle.

The dance said, "This pollen is not far from our hive."

The bee did the round dance several times on the same spot of the comb. Honeybee kept her antennae close to the dancer's body to smell the strong, rich odor of the pollen. When the dancer stopped, Honeybee tasted the pollen.

She wanted to find that pollen and rushed out of the hive. *Zoom!* She buzzed into the air, flying low where there was no breeze to blow her about.

Honeybee flew this way and that, a few feet from the hive. For this work she needed her eyes, all five of them. She saw something yellow in the green grass, flew close, and hovered while she smelled it. This was not the odor she was hunting for. This was a daisy.

She flew on, across a patch of red clover which she did not see because she was blind to red, like all honeybees.

She met many butterflies and bumblebees.

On she flew, buzzing this way and that, smelling the flowers she passed by. None of their pollen had the right odor.

She came to a creek and caught a whiff of the odor she was hunting. There were only two kinds of flowers growing there, blue iris and yellow pussy willow.

She hovered over the pussy willow.

Here it was—the same odor as the dancing bee's pollen! The stamens of the flowers were full of pollen dust.

GOLD DUST

Honeybee hovered over the stamens. As she beat her wings, the air dusted the pollen all over her. It caught on the hairs of her body and on the hairs of her legs.

All the time she hovered, her front and middle legs were brushing the pollen back to her hind legs. It stuck there, among the stiff hairs of her pollen basket.

She flew to other pussy willows and gathered pollen. She did not know she was helping the pussy willows, too. By mixing the pollen of one with another, she was fertilizing them so that their seeds would grow.

Her pollen baskets got so full that her hind legs looked like flowers themselves.

She stopped at the creek. Carefully she stepped to the sandy edge and sucked up the cool water.

She saw the water come up and catch a bee and take it out into the stream. She saw a robber fly chasing a bee and a kingbird chasing the robber fly. She saw dragonflies darting above the water.

The creek was a place of danger.

After they had gone, Honeybee flew up from the water with her heavy load.

She remembered to look for the apple tree and the fence. Then she saw her own white hive amid the grass and the dandelions.

At the entrance she touched antennae with a guard, who let her pass. When she got to the combs, she crossed her legs and brushed off the pollen into a cell.

While Honeybee was in the field, six new larvae had hatched—six female larvae that were just like Honeybee when she hatched.

But these larvae were in very large, long cells, and the nurses would feed them nothing

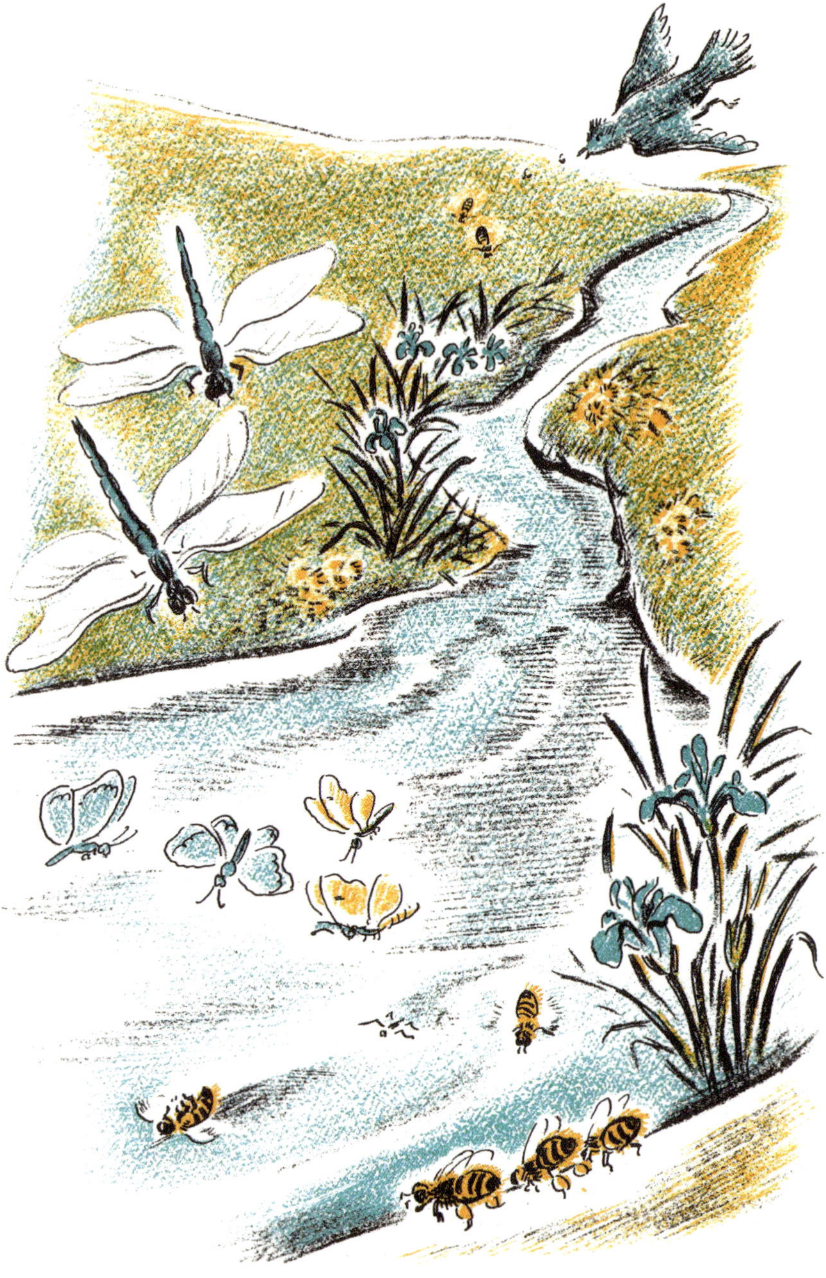

but royal jelly, for they were to grow into queens.

The hive would soon need a new queen, for the old one would be leaving, as old queens do in spring.

FLYING BY THE SUN

Every day Honeybee went to the fields.

Once she was out scouting for flowers in bloom and flew far from the hive. She flew toward the sun, stopping now and then to smell a flower. She had never been so far, but she watched the sun to keep it straight ahead.

This was easy to do, with her great eyes on top of her head where they could watch the sky.

She had gone a mile when she found a field of white clover. She buried her head in the blossoms and sucked up nectar until she filled her honey sac. She flew back, keeping the sun behind her. This was just opposite from the way she had come, so she knew the direction was right.

It was a long, long flight, carrying a load that weighed almost as much as she did.

WAGGING DANCE

The minute she got home, Honeybee began to dance. The white clover was too far away for her to tell the other bees with the round dance. So she danced the wagging dance.

She wagged her body from side to side while she ran a little way straight up the comb, in the direction of the sun.

She stopped and turned all the way around to the left, making a circle. Then she again ran straight up the comb, wagging from side to side. She turned to the right this time and made another circle.

She did these runs and circles over and over before she stopped.

The field bees understood. Running straight up on the comb said, "Fly with the sun straight ahead of you." The number of circles in the dance told them how far the nectar was from the hive.

The field bees kept smelling the perfume of the clover nectar on Honeybee's body. When she stopped, each took a taste of the nectar from her tongue. They rushed out to find the clover.

Honeybee gave her nectar to a house bee, to finish changing it into honey, as she herself had done when she was a house bee. Then she, too, rushed out to the clover.

SWARM HOLIDAY

One morning very early, the nurse bees sealed the cells of the queen larvae. When they awakened, six young queens would be born.

This news spread quickly through the hive. The bees buzzed excitedly, for they knew that now the old queen would leave and that many of them would go, too. It was swarm time, time to find a new home.

The queen stopped laying eggs. None of the field bees went out. None of the house bees worked. Only the nurse bees worked so that the larvae would not starve. Swarm time was the great holiday of their year.

The bees raced and buzzed about the hive, knocking many young bees off the comb. They tore open the honey cells.

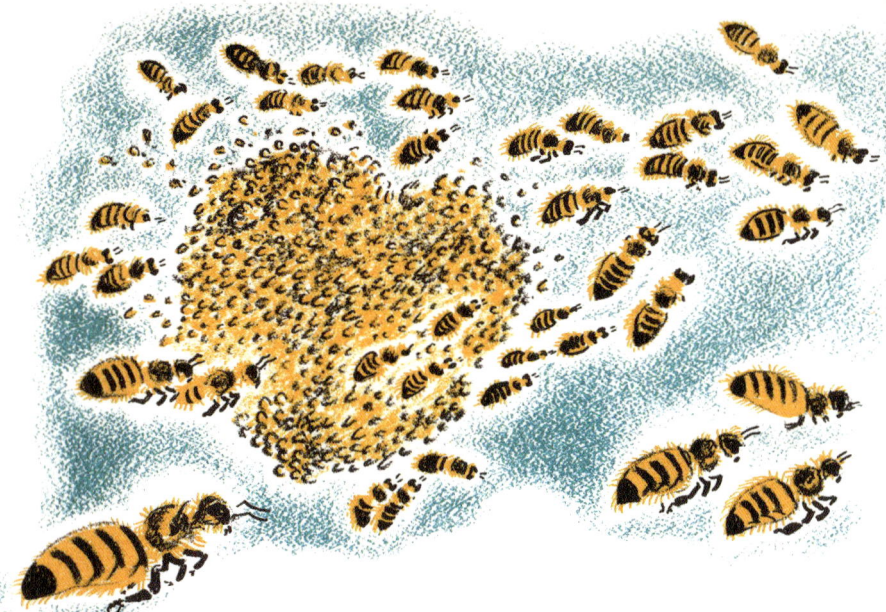

Honeybee stuffed herself with honey.

Then she raced to the opening with all the other bees. They pushed and squirmed. They tumbled over each other as they came out onto the landing board.

Honeybee darted into the air among the thousands and thousands of other bees circling above the hive. The sunlight danced on their lacy wings and black and yellow bodies.

Bees and more bees tumbled out of the hive. Among them came the queen. It was the first time she had seen the sunshine since her mating flight a year ago. She flew into the air, and thousands of bees closed around her. The swarm circled above the hive until about half the bees had come out.

Honeybee smelled a signal and followed it away from the swarm. Scout bees were sending out scent signals from a new home they had found in the hollow trunk of an oak tree. They were letting perfume out of the scent organs on the tips of their abdomens and fanning. The perfume was floating back toward the swarm.

More and more bees in the swarm smelled the scent. Soon the whole swarm was following the signals. The mass of bees sailed through the air in a huge ball with the queen inside.

When the swarm reached the oak tree, the ball broke up into a rain of bees and they landed, with the queen.

They poured into the hollow trunk.

Honeybee was not with them. A little gust of air had blown away the scent she was following, and she had lost the swarm.

But Honeybee herself was not lost. She flew back to the hive and stayed there, as if she had changed her mind about swarming.

BATTLE OF QUEENS

A week later the first two young queens came out of their cells at the same time. They rushed at each other, for they knew that the hive could have only one queen.

The two queens fought. Each grabbed the other's antennae and held tightly. Honeybee and other workers pulled them apart and made circles of guards around each queen.

One queen was fierce and broke away.

She ran to the other circle, where Honeybee was helping to guard the other queen. She climbed over Honeybee, who backed away. The fierce queen stabbed the other queen with her sting and killed her.

Then she ran over the comb, hunting for the other queen cells and making a high piping sound. She found the cells. She bit them open and stung the young queens to death.

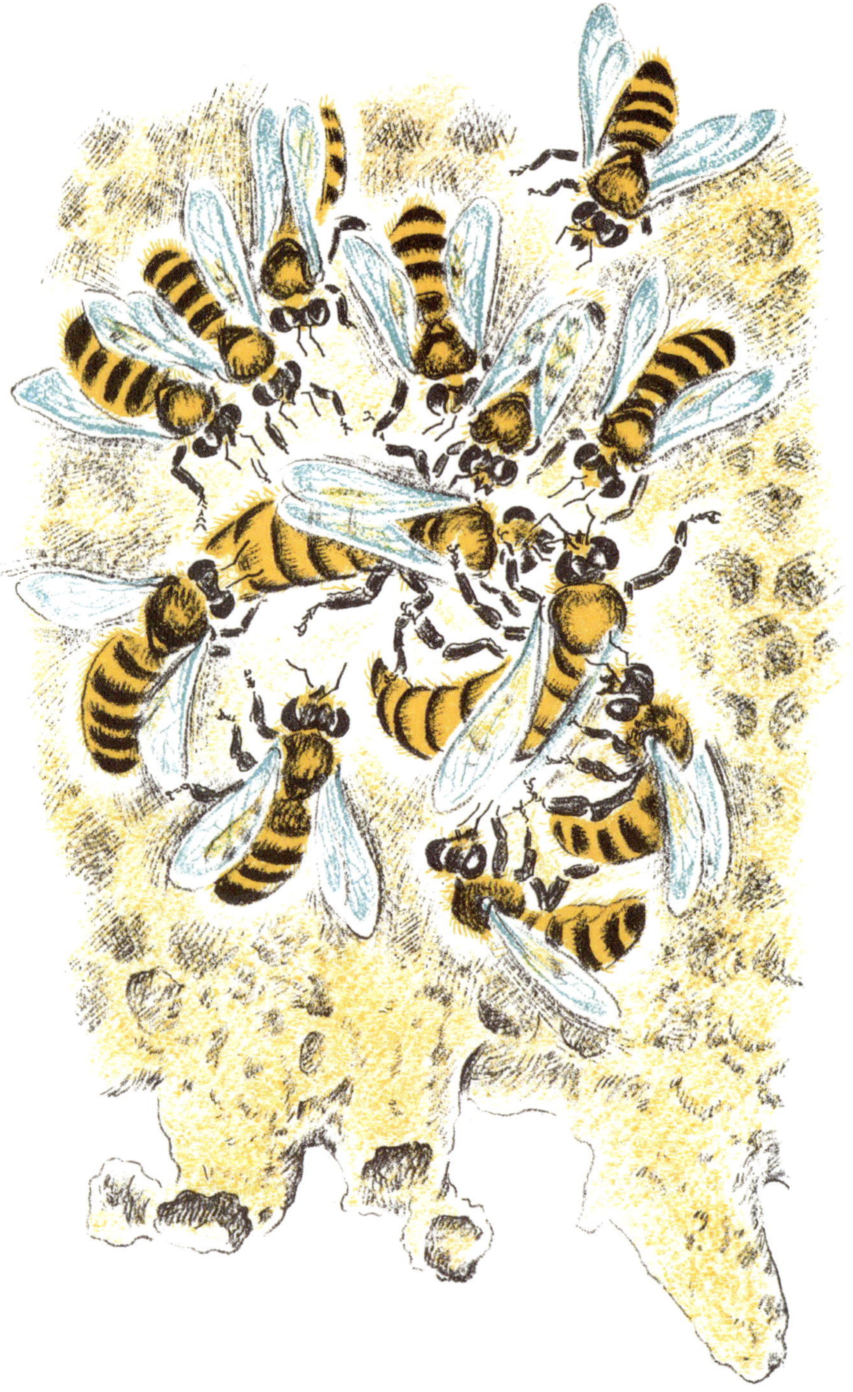

The workers came to their new queen and began cleaning and feeding her.

She ate hungrily. Now she would become the gentle mother of the hive. She had a long life of three years ahead of her, a busy life of egg-laying. But first she must mate to fertilize her eggs.

MATING FLIGHT

A few mornings later, Honeybee felt too tired to go to the fields. She had been working too hard. Soon she would die. So she stayed on the landing board in the warm sunshine.

The young golden queen came out of the hive and sprang into the air. Her strong wings carried her in a circling flight, up and up, toward the sun.

Some of the drones flying above the hive saw the queen and followed.

Faster and higher, higher and faster, circled the queen. She saw the drones below, following her. Higher and higher she circled, as high as the top of the apple tree. Three drones flew past the others.

Still toward the sun, the queen circled. One drone zoomed past the others. He caught up with the queen.

They faced each other, their wings beating and shining in the sunlight. They came together and held to each other as they flew, and he put a liquid from his body, called sperm, into a sac on the queen's body.

Down, down they sailed to the ground. The drone fell into the grass and died, as do all drones while mating.

Two days later, the young queen began to lay eggs. When she wanted the eggs to hatch into workers, or female bees, she fertilized them with the sperm. But when she wanted the eggs to hatch into drones, or male bees, she did not fertilize them.

The life of the drone who had died was in the sperm. So his life became part of the new lives of thousands of honeybees.

The hive needed these new lives, for it had lost thousands of bees in the swarm. Also, Honeybee and many other workers were getting old and would soon die.

The young queen laid so many eggs that the hive was full of young workers all summer. But when fall came and the flowers stopped blooming, she stopped laying eggs. The hive would not need many workers in the winter. The youngest workers would live until spring, because they would be resting.

So now the young honeybees gathered bee glue from leaf buds and sealed every crack to keep out the cold.

With plenty of food and a warm hive, the new honeybees were ready for winter.

THE END

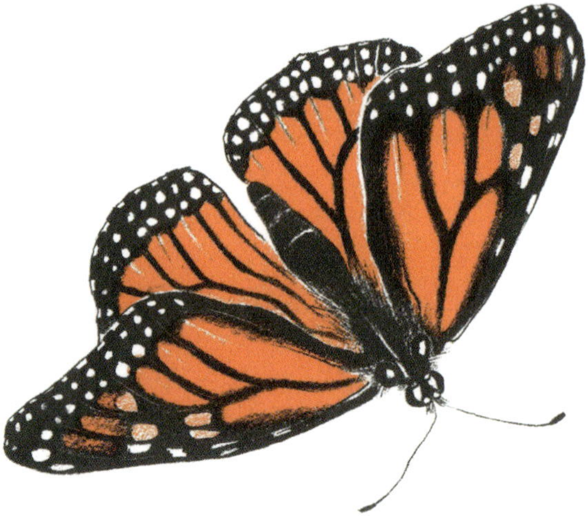

MONARCH BUTTERFLY

by Marion W. Marcher

illustrated by Barbara Latham

Originally published in 1954

FOREWORD

The monarch butterfly may be seen throughout the United States and southern Canada. It is often called the milkweed butterfly because, in the caterpillar stage, it feeds upon milkweed leaves. Its scientific name is *Danaus plexippus*.

The monarch's capacity for travel is phenomenal. Its long autumn migrations to the south, often in immense numbers, have made it perhaps the most famous of all butterflies. Stray individuals frequently cross the oceans, probably assisted by hitchhiking on ships.

The smaller viceroy butterfly has similar color and markings, but it can be distinguished by the thin black line across its hind wings.

Butterflies belong to the same order (Lepidoptera) of insects as moths, and both go through the same stages of development, but the adults differ in appearance and behavior. Butterfly antennae are slender and end in a swelling or knob; moth antennae very rarely end in knobs, and some are feathered. Most butterflies fly by day and rest with the wings raised upright. Most moths fly by night and rest with the wings horizontal or depressed.

I wish to thank Mr. Austin H. Clark of the Department of Zoology, Smithsonian Institution, for checking the scientific data.

Marion W. Marcher
Milwaukee, Wisconsin

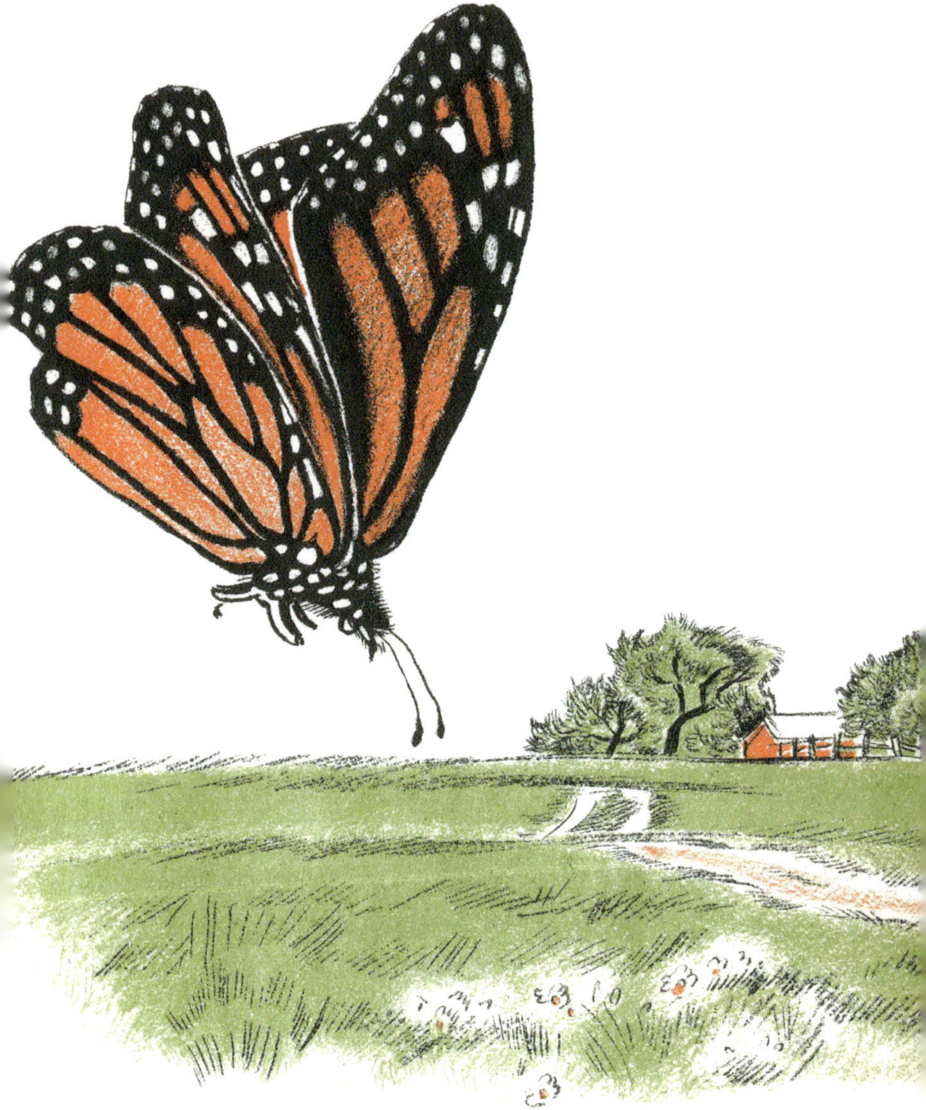

SPRING FLIGHT NORTH

In the clear spring air, Monarch Butterfly was winging her way across the country. She was an old and faded beauty of the Royal Family of butterflies, but her orange-brown wings still flashed in the sunshine.

All winter she had been in the warm southland, like the robins. Now she was going north. She flew slowly and not far from the ground, hunting for milkweed as she went along.

Over fields and along hillsides she flew and then down through valleys. No river or lake was too wide for her to cross. She hurried over the water, flying above people in boats.

As soon as she came overland again, she flew lower and slower. She stopped moving her wings and coasted along in twirly glides. She dipped her wings on one side then on the other. This made her tilt from one side to the other, but it steered her wherever she wished to go.

The sky became cloudy. Rain began to fall. Monarch found a tree and fluttered down upon a large branch.

Spreading her wings to balance herself, she stepped to the underside of the branch. Here she had a roof to get under, out of the cold rain.

She fastened into this roof using the tiny hooks that were on the ends of her legs and hung downward. She folded her wings together and rested. Rain splashed on her roof all night long, but she was dry.

The next morning, the sun shone. Monarch smelled delicious flower nectar. She smelled it with the two thread-like feelers on her head. These were her antennae. She smelled the nectar with another pair of feelers, too. They were feathery little palps on either side of her tongue.

Soon she saw where the delicious smell came from—clover blossoms. She chose a nice big one. The moment she alighted on it, she tasted it with her feet. Her wings quivered.

Quickly she uncoiled her tube-like tongue and pushed it deep into the blossom. As she sipped the sweet nectar, her wings folded slowly in contentment.

All around her other flying insects were stopping to enjoy this sweet food, too.

After Monarch had sipped nectar from several clover blossoms, she coiled her tongue back neatly in place between the two feathery palps. She opened and closed her orange-brown wings to warm them in the sun while she rested. Then she took off again flying northward.

TENDER LEAVES AND TINY EGGS

Monarch flew on and on, always northward, hurrying over bare places and lingering in flowery places to eat.

One morning, as she fluttered along a country roadside, she smelled something very good. It was milkweed leaves.

She bumped against a leaf gently and could tell it was soft just by feeling it with the fine hairs on her body. She bumped against another leaf that was still softer. She stopped and felt it with the ends of her antennae as if they were fingers. Then she alighted on the leaf.

Monarch bent herself in a curve over the edge and pressed the end of her body lightly against the underside of the leaf. There she laid a tiny egg.

It was pale green, moist, and very sticky. But in the hot sunshine, it dried quickly and stuck tight to the leaf. So sunny days were the only kind for Monarch to do her egg-laying.

Farther up the road, she came to another small young milkweed with tender leaves. As she circled it, her wings hit the leaves, shaking them. Some black ants on them scurried off, frightened. She laid three eggs, each on its own leaf. After she flew away, the ants came back and ate two of the eggs. A tiny red mite found the other egg and sucked out the wee drop of liquid in it.

But Monarch had many more eggs to lay. So now she traveled northward more slowly. Sometimes she put her egg on the upper side of a leaf instead of the under. Always, though, she let each egg have a whole tender leaf to itself. And always she laid her eggs on milkweed.

One day, weeks later, she found milkweed in a backyard where some girls and boys were playing. They saw a large butterfly with faded orange-brown wings. But they did not know she had flown many hundreds of miles and that now she was going to lay her last eggs on the milkweed plant in their yard.

Monarch laid the last egg of all on the upper side of a top leaf.

All that was left for her to do was to rest in

143

sunny places these first days of summer. In a meadow, she found the sweet-smelling pink blossoms of milkweed. Their nectar was her favorite food now. And this was as far north as she ever got, for her wings were worn. They were colorless in the parts where the scales had worn off. So she could not fly well anymore.

At last, she rested on one blossom all day, too weak to move. When she could cling to it no longer, she fell into the grass.

Monarch was over eight months old. She had lived longer than most kinds of butterflies. She had laid almost three hundred eggs.

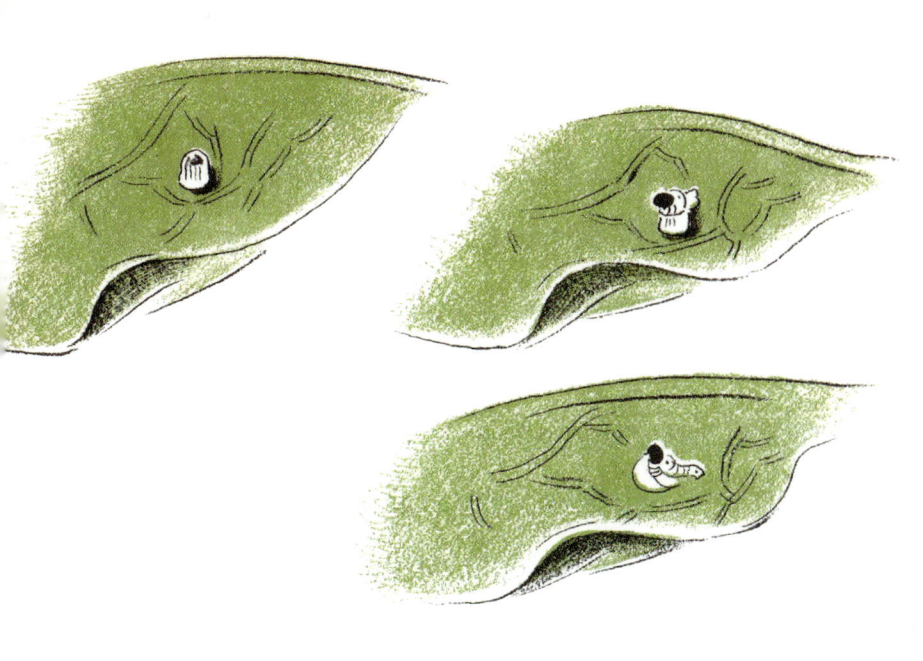

CATERPILLAR'S FIRST DAYS

Only a day after Monarch's last egg was laid, life began stirring inside it. The next day, this life became Monarch Caterpillar moving about in the shell.

On the third day, she bit out a hole in the top of it. She pushed up her shiny black head. Out she climbed and crawled down the side of the shell, and then turned right around and ate it up.

Her body was gray-green, which made it hard for her enemies to see her against the gray-green milkweed leaf.

She crawled around to the underside. Here in the shadow, she was safer still. This good leaf that her mother had found for her was tender and juicy to eat. Caterpillar was only an eighth of an inch long, but she was built to be a big eater.

Her tiny jaws were like sharp nippers. She nipped away at the leaf, leaving a hole everywhere she ate.

She ate and rested and ate again. She ate and ate all that first day and the next. She ate so much and grew so fast that her skin got too tight for her. Before she could grow any more, she had to get out of it.

The way she did this was to spin a silk carpet first. She had some sticky silk thread that came from a spinneret on her lower lip. So she crawled forward on the leaf, moving her head from side to side while laying down the silk thread.

Forward and back again she went until she had a carpet the same length as her body.

She had hooks on some of her legs, and so she hooked herself close to the carpet. Then she waited for hours and hours. Meanwhile, a new skin was forming under the old skin, which began to give way. She worked to split the old skin.

She puffed herself up by breathing in extra air through the tubes along her sides. Suddenly, the old skin split near her head.

With her hooks she held tight to the carpet and strained and struggled. At last, she pulled out of her old skin, headfirst.

Very tired now, she rested while her new skin dried and toughened. Then she ate the old skin she was done with, just as she had eaten her eggshell.

LITTLE BLUFFER

Caterpillar's new skin was roomier, but she ate so much milkweed and grew so fast that soon she could not stretch her skin any bigger. Again she had to get out of it—and again and again.

Caterpillar had a slow, creeping way of life. She had eight pairs of legs to do this with, though.

Five stubby pairs were for props to hold up her long body. These prop legs were the ones with the hooks. At night she rested with them hooked into the underside of a leaf.

The other three pairs of legs were near her head. She used them to crawl and to feel her way around, for she could not see very well with her small eyes.

As she grew larger, her skin became beautifully marked with rings of yellow, black, and white. Near her head she grew a pair of black horns that were like little whips. She had another pair at the end of her body, but they were shorter.

Once when she was eating on the top side of a leaf, two flies alighted on it. They tried to lay eggs on her. But she jerked her head and lashed those horns at them so fiercely that they flew away. If they had laid their eggs, little maggots would have hatched and fed on Caterpillar's body and killed her.

The sun became so warm that she crawled along to the underside of another leaf. There she rested in the cool shade. Suddenly a grasshopper thumped down on top of the leaf, and it bounced up and down.

Caterpillar curled up into a ball, frightened, and fell down into the grass. She pretended to be dead.

After a long while, she began crawling as fast as she could over to a new milkweed plant.

A song sparrow flew over and saw her, but he did not eat her. This was because the milkweed that tasted so good to her made her flesh too bitter for birds to like.

An ichneumon (ik-NOO-mahn) wasp saw her crawling along. This insect was flying around the grasstops, trailing thin black legs behind. She, like the flies, wanted to lay her eggs on a caterpillar so that the young ichneumon larvae could eat it when they hatched.

The wasp flew closer. She saw that this was the wrong kind of caterpillar for her eggs. She laid them on other kinds, especially certain moth caterpillars that destroyed trees.

So the ichneumon wasp flew away, and Caterpillar crawled up the milkweed.

153

LITTLE ACROBAT

So Caterpillar lived on. When she became two weeks old, she was full-grown at two inches long. She had changed to a larger skin four times and always in the same way.

The next time she shed her skin, though, it was the last time and very different.

She crawled down her milkweed stalk and away through the grass. She traveled for hours. Often she stopped and stretched upward to feel for a higher place. At last, she found one when she came to a low log fence. She climbed up and up until she reached the under part of the top log. It was a dark, cool, and quiet hiding place.

So she spun another silk carpet. But this time she made a small bump of silk on the carpet. Then she put her hind hooks into the bump and let herself hang downward. She was like a circus acrobat hanging by the feet.

All day and all night, Caterpillar hung there. All night she kept trying to split her old skin. She breathed in more air. She even pumped her blood in a certain way to puff herself up.

In the morning, she split the skin near her head. She stretched up and down to work the skin off from her head to her tail.

And as the skin came off, all the feet came with it! Not only that, but the head and horns, too!

What was left was not Caterpillar anymore. It was a chrysalis (KRIS-uh-lis), or pupa (PEW-pa), forming.

While it was forming, though, it too was like a little acrobat. It had to do the hardest trick of all so far.

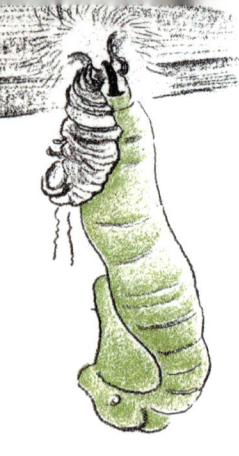

Now all the skin was in a bunch near the tail end. The trick was to slip this old skin off over the tail feet without losing hold of the ceiling and falling to the ground.

The little "acrobat" did the trick with some hooks of its own. They were covered over by the old skin that was bunched up at the tail end. So, first, the "acrobat" drew these hooks out and then reached with them up and around the bunched skin.

She reached higher and higher. There! The new hooks went into the silk bump. With a turn and a twist, the old skin fell to the ground. There the chrysalis hung, very still now. It was soft and moist. As the skin dried, the chrysalis became a shining green. Near the top were gold dots in a row, and lower down were some more. It was like a little greenhouse trimmed in gold.

Inside it, a new monarch butterfly was forming.

Many organs in the chrysalis' body were now dissolving into liquid. Then, from the liquid, new organs began to grow—those that a butterfly would need.

Some parts of the body were being made over. The chewing mouth was changing into a sucking mouth. Thousands of tiny eyes were growing in a cluster to make two great eyes. Tiny folds in the skin were growing into wings.

A few parts hardly changed at all.

THE NEW MONARCH

Rain fell, but the waxy skin of the chrysalis was waterproof.

Several days passed. The shining green faded. On the seventh day, the orange-brown and black marking of butterfly wings showed through the thin skin of the chrysalis.

Early the next morning, the new butterfly tried to break out. It was hard work.

She puffed herself up—more and more and some more. Then *click!* The chrysalis split open a little. It was in the lower part where her head was, for she was hanging head-down inside the chrysalis.

She pushed gently against the split. It opened wider and wider then wide enough so she could get a foot out and reach for a foothold. Next, she freed another foot and held on with it, too.

Now she drew her body downward until all of her was outside, clinging to the skin as she hung there upside down.

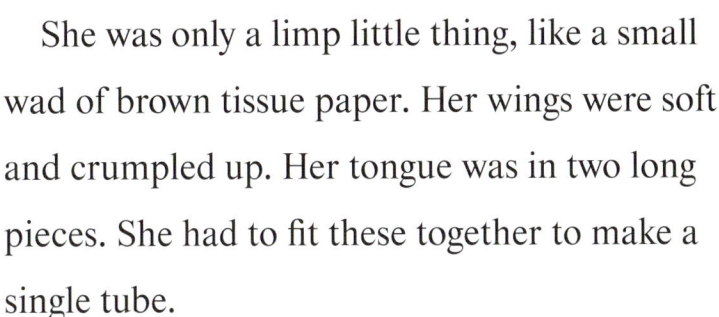

She was only a limp little thing, like a small wad of brown tissue paper. Her wings were soft and crumpled up. Her tongue was in two long pieces. She had to fit these together to make a single tube.

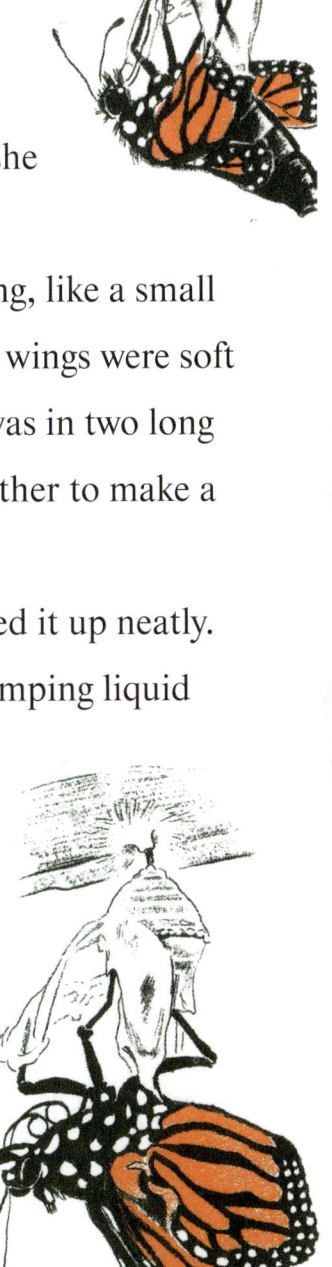

At once she did this and coiled it up neatly. At the same time, she began pumping liquid from her body into the veins of her wings. As these veins filled, they stiffened and became a strong framework inside her wings, which began to unfold and stand up firm. She fanned them back and forth slowly to to dry them and smooth out the crinkles. Now, at last, her wings spread out to their full size.

Monarch Butterfly was here again! She was already in her full beauty and full growth. She was as large now as she would ever be.

The colors on her wings were made of thousands of dust-fine scales. Each scale lay lapped over the next one, like the scales on a fish or shingles on a roof. And these tiny scales could shed the rain.

Monarch walked on up to the top of the log

and waited a moment before making her trial flight. She looked around her, moving her head this way and that, and saw grass and flowers nearby.

Slowly now, she moved her wings up and down, trying them, up and down, then a quick flutter and off she flew. Her orange-brown wings flashed almost red in the bright summer sunshine.

She was not yet strong enough to fly high or long. Soon she circled slowly downward to some lovely white clover.

The new monarch took her first taste of nectar. She tried another blossom and another. She tasted thistle, alfalfa, milkweed, and many others. She liked them all.

POCKETS OF PERFUME

As she flew along, she came to a garden.

Several cabbage butterflies were dancing together in the sunshine. They liked to dance this way and kept it up for a long time.

Down at a pool, some other butterflies were sipping the water. Monarch fluttered down for a drink, too. A small pearl crescent butterfly dashed at her, and she flew away rather than fight with it.

She settled on a white aster. There a buckeye butterfly came at her angrily. But peaceable Monarch sailed slowly away from him and then circled down to the pool.

She sipped her drink beside a painted lady butterfly, which did not bother her.

Monarch lingered a while in the flower garden and then wandered onward.

A few days later, when a strong breeze was

163

blowing, she flew bravely into it and let it toss her about. Up and up she flew. Then she held her wings steady and coasted downward on an air current. She dipped her wings to one side and then to the other to steer herself.

She was now a perfect flier.

She became very bold with the small birds. In a city park one morning, she was sipping honeysuckle. A hummingbird came hovering over her flower. She turned and faced her whir of wings. Quick as a flash, she darted straight at him.

He flew away frightened. For a while longer, Monarch snapped her wings in excitement and then returned to her flower.

Her next adventure was along a creek. A new scent came to her on the breeze. She liked it so much that she followed it. It led her straight to another monarch.

He was feeding on some goldenrod nectar.

His wings were like hers except that he had a small black patch on each hind wing. These were pockets of perfume.

The nearer she got to it, the more she liked it. So she alighted on a spray of goldenrod near him. He fluttered over and perched beside her. She chose him for her mate. Away the two of them flew.

After some days, Monarch lost sight of her

mate. Weeks passed. Every few days as she flew about, she found milkweed and laid eggs.

Just as her mother had done in the spring, she chose a tender leaf for each egg. And she, too, laid almost three hundred.

FALL FLIGHT SOUTH

One afternoon in early autumn, she came to a field where many monarch butterflies were feeding. She alighted on a thistle beside two of them. Soon, another monarch joined them, and then another, and another.

All over the field, on the plants, the butterflies were gathering into little groups and then fluttering away to other groups. These seemed very restless.

Monarch left the group on the thistle and settled herself on a bush for the night. Other monarchs were resting there, too. And before dark many more joined them. On nearby trees there were still other groups resting among the lower branches.

The next morning, Monarch took off southward. Beyond her some monarchs were already in the sky. Behind her others were coming from the field. They drew together, in a small company, as they flew southward.

Other companies of monarchs joined them. Farther on, more companies joined in until there were thousands and thousands of monarch butterflies all flying together, very high, making an orange-brown cloud in the sky.

Last fall, Monarch's grandmother had flown south in a great cloud like this. But Monarch had never been south, nor had any of the butterflies who were with her. Yet they all knew the way.

They drifted restfully on the wind when it blew south. When the wind blew in other directions, they beat their wings steadily against it. Whole days they flew, and when they were tired, they all rested together.

They flew over towns and cities and farmlands, across highways and telegraph wires, along river valleys, and over lakes and plains. After many days, and many hundreds of miles, they were in the south.

They flew straight toward the pine trees that the monarchs came to every fall. These trees were their winter hotel.

On sunny days Monarch and the other butterflies fluttered through the southern gardens. But at sundown each one returned to its own home tree.

On cold days Monarch stayed huddled with the others in her pine tree.

When spring came, every one of them left for the north. But now each butterfly traveled alone.

So, as her mother had done the spring before, Monarch winged her way northward. She would not live to fly all the way north, for she was already nearly eight months old. But, old as she was, Monarch began the journey, her orange-brown wings flashing in the sunshine.

171

HOW TO RAISE A MONARCH BUTTERFLY

Find a stalk of milkweed that has a leaf with a caterpillar or an egg on it. Pull up the stalk and put it in a jar partly full of water. Over the mouth of the jar, tie a cloth with a slit in it for the stem so your caterpillar cannot get into the water and drown.

Add fresh milkweed often for it to feed on. Never take it off the leaf if it is not feeding. It may be shedding its skin and cannot spin a second carpet.

When your caterpillar is nearly full grown, put in the jar a stalk of fresh milkweed that has large sturdy leaves on it. Tie mosquito netting over this loosely so it will not touch your caterpillar when it becomes a "little acrobat." (That will be when it hangs head-downward on the underside of a leaf.) Then take the netting off very carefully, so as not to shake the "acrobat."

Watch and watch now. When your caterpillar's "horns" look limp and twisted, watch for it to change to a chrysalis. This usually happens early in the morning and takes only about five minutes. So plan to get up very early.

Let nothing touch the chrysalis.

When you can see wings plainly through the chrysalis skin, watch for the butterfly. Usually, it comes out early in the morning and takes only five minutes, too. Don't miss this wonderful show either.

After your butterfly's wings are dry, which takes about three hours, let it walk onto your finger and take off from *there!*

Let it be free to fly away, hundreds and hundreds of miles, and live the life you read about in this book.

THE END

MORE BOOKS FROM THE GOOD AND THE BEAUTIFUL LIBRARY

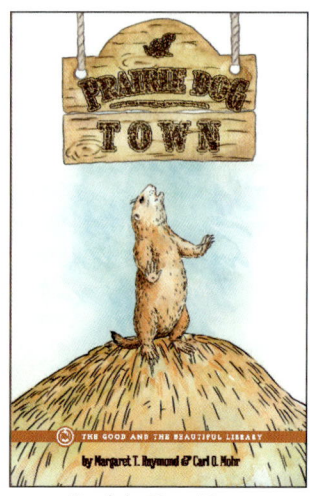

Prairie Dog Town
by Margaret T. Raymond &
Carl O. Mohr

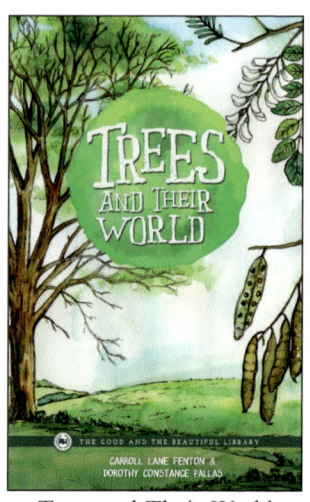

Trees and Their World
by Carroll Lane Fenton &
Dorothy Constance Pallas

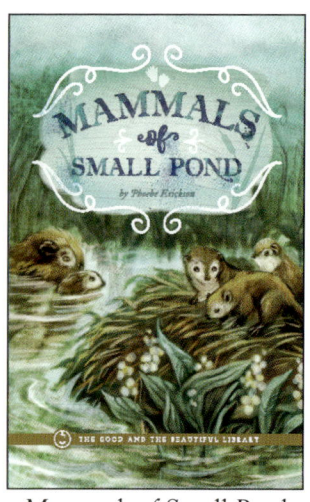

Mammals of Small Pond
by Phoebe Erickson

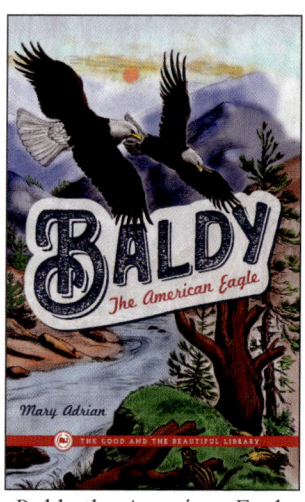

Baldy the American Eagle
by Mary Adrian

GOODANDBEAUTIFUL.COM